Markus Mueck and Ingolf Karls
Networking Vehicles to Everything

Markus Mueck and Ingolf Karls

Networking Vehicles to Everything

—

Evolving Automotive Solutions

DE
—
G
PRESS

ISBN 978-1-5015-1572-9
e-ISBN (PDF) 978-1-5015-0724-3
e-ISBN (EPUB) 978-1-5015-0720-5

Library of Congress Cataloging-in-Publication Data
A CIP catalog record for this book has been applied for at the Library of Congress.

Bibliographic information published by the Deutsche Nationalbibliothek
The Deutsche Nationalbibliothek lists this publication in the Deutsche Nationalbibliografie;
detailed bibliographic data are available on the Internet at http://dnb.dnb.de.

© 2018 Markus Mueck and Ingolf Karls
Published by Walter de Gruyter Inc., Boston/Berlin
Printing and binding: CPI books GmbH, Leck
♾ Printed on acid-free paper
Printed in Germany

www.degruyter.com

Acknowledgments

It is with pleasure that we thank all the collaborators and colleagues at Intel's Automated Driving Group (ADG), the Communications and Devices Group, and the Next Generation and Standards team who worked along with us during the years that have led to this book. We cannot thank them one by one, but their contributions were very helpful to get to this point. Finally, we express our sincere gratitude to De Gruyter for giving us the opportunity to complete this project. In particular, we are very grateful to Jeffrey M. Pepper, who believed in this project from the very beginning, and our thanks also go to Jaya Dalal and Natalie Jones, our production project team, who were very kind and very patient.

Munich, November 2017

Ingolf Karls
Markus-Dominik Mueck

DOI 10.1515/9781501507243-201

Contents

Chapter 1: Introduction —— 1
1.1   V2X objectives —— 6
1.2   V2X and D2D networking and connectivity —— 9
1.3   Technical challenges for V2X —— 10
1.3.1   Sensors —— 10
1.3.2   Computing —— 12
1.3.3   Networking —— 13
1.4   Society, ethics and politics —— 17
1.5   Outline —— 23

Chapter 2: Applications and Use Cases —— 27
2.1   Use cases up to 2005 —— 29
2.2   Between 2005 and 2011 —— 32
2.3   Use cases since 2011 —— 38
2.4   Conclusions —— 50

Chapter 3: V2X Requirements, Standards, and Regulations —— 59
3.1   Requirements —— 61
3.1.1   Sensors —— 65
3.1.2   Communications —— 66
3.1.3   Dynamic high-definition maps —— 69
3.1.4   Over-the-air updates —— 71
3.1.5   In-vehicle Infotainment —— 71
3.2   V2X networking and connectivity standards —— 73
3.3   V2X networking and connectivity regulation —— 79
3.4   Conclusions —— 85

Chapter 4: Technologies —— 89
4.1   Sensing —— 91
4.2   Computing —— 96
4.3   Communications —— 99
4.4   Software —— 105
4.5   HAD maps —— 108
4.6   Functional safety —— 111
4.7   Conclusions —— 113

Chapter 5: V2X networking and connectivity —— 121
5.1   IEEE 802.11p-based DSRC and ITS-G5 —— 121
5.2   LTE and 5G NR V2X —— 128
5.3   C-V2X Phase 1 where 3GPP works on LTE Vehicular Services —— 131
5.3.1   Vehicle-to-vehicle (V2V) Communications —— 137
5.3.2   Vehicle-to-pedestrian (V2P, P2V) Communications —— 138
5.3.3   Vehicle-to-infrastructure (V2I, I2V) Communications —— 138
5.3.4   Vehicle-to-network Communications (V2N, N2V) —— 139

5.3.5   Synchronization —— **140**
5.3.6   Location Determination —— **142**
5.4   C-V2X evolution toward Phase 2 and beyond —— **143**

Chapter 6: Infotainment —— 151
6.1   Infotainment —— **152**
6.2   Telematics and control —— **156**
6.3   In-vehicle connectivity and networking —— **158**
6.4   Conclusions —— **162**

Chapter 7: Software Reconfiguration —— 167
7.1   Issues to Be Addressed by Software Reconfiguration —— **168**
7.1.1   Problem statement 1: How to transfer and install radio software
           components to a target platform in a secure way —— **168**
7.1.2   Problem statement 2: How to enable a user to access new software
           components —— **168**
7.1.3   Problem statement 3: How to deal with device certification in the
           context of novel radio software components —— **169**
7.1.4   Problem statement 4: How to achieve software portability and
           execution efficiency —— **169**
7.1.5   Problem statement 5: How to enable a gradual evolution toward
           software reconfigurability —— **169**
7.2   The Regulation Framework: Relationship Between Software
           Reconfigurability and the Radio Equipment Directive —— **169**
7.3   Mobile Device Reconfiguration Classes: How the Level Of Reconfigurability
           Will Grow Over Time —— **172**
7.4   The ETSI Reconfiguration Ecosystem —— **174**
7.5   Code Efficiency and Portability Through the Radio Virtual Machine —— **180**
7.6   Multi-Radio Interface (MURI) —— **184**
7.7   RRS Software Reconfigurability For Hardware Updates —— **186**
7.8   The ETSI Software Reconfiguration Solution in Comparison to Other
           Alternatives —— **187**
7.9   ETSI Security Framework for Software Reconfiguration —— **188**
7.10   Responsibility Management —— **190**
7.11   Conclusion —— **193**

Chapter 8: Outlook —— 195

Appendix A —— 203
1.1   List of acronyms —— **203**

Index —— 251

Preface

Two rising tidal waves of automated and autonomous driving vehicles and the beginnings of the next generation of wireless communications (5G), both game changing technologies for global vehicle ecosystem stakeholders, have recently come center stage. V2X networking and connectivity technologies span several different research, development, and standardization and regulation fields ranging from sensors, computing, communications, and functional safety to software reconfiguration. It is of utmost importance to understand the contributions of V2X networking and connectivity on many real-world transport and travel use cases. Among these use cases, those that aim to ensure the abidance of traffic laws and road safety are extremely challenging. Some that are linked with new business models, and which try to change the ways in which goods and people are transported, are already feasible.

It is apparent that fact-finding and forward looking insights into V2X networking and connectivity with regards to automated and autonomous driving vehicles are valuable resources for those who are part of the rapidly changing vehicle ecosystem. *Networking Vehicles to Everything* is the work of two experts in the field of wireless communications and control engineering. The book presents an immense range of social, ethical, and technical aspects such as standardization, regulation, computing, communications, networking, and safety and security. It successfully covers many V2X networking and connectivity use cases stemming from research and development that have been researched over ten years.

Moreover, we describe, in a comprehensible fashion, a number of interleaved standardization and regulation activities driven by the requirements of advancing automated and autonomous driving, comparing European, U.S., and Asian visions. Along the way, the book provides an insightful impression of the major automated and autonomous driving vehicle technologies, like sensors, dynamic high-definition maps, over-the-air-updates, and in-vehicle infotainment, in relationship with communications. In particular, one of the unique points of this book is the presentation of the state of most current potential solutions for V2X networking and connectivity encompassing both DSRC-oriented and LTE-oriented perspectives.

The book presents the principles of emerging global V2X networking and connectivity solutions, in particular it illustrates how IEEE 802.11p-based DSRC and ITS-G5 and LTE and 5G NR V2X apply to the most relevant envisioned V2X networking and connectivity use cases for automated and autonomous driving. An in-depth analysis of the in-vehicle infotainment use case in association with the vehicle networking paradigms and technologies currently in use for wireline communications is provided. Key challenges of integrating V2X networking and connectivity with vehicle systems for automated and autonomous driving taking into account the very different product life cycles of vehicles and communications hardware and software are

DOI 10.1515/9781501507243-202

discussed. We elaborate on the most important feasible solutions of software reconfigurable radio systems.

This book by no means covers all aspects of V2X networking and connectivity and automated and autonomous driving vehicles. But it represents an excellent reference for automated and autonomous driving solutions as of today for those who are working in the field. We are positive that all readers will appreciate it. The book attempts to look intently at the current hype on the evolution of V2X and automated and autonomous driving from a perspective encompassing advances in wireless communications technologies that we came across during our research in the last few years. Our hope is that this book, with our specific conclusions and outcomes, will be used as a starting point and a useful comparative reference.

Chapter 1
Introduction

Automated and autonomous driving will change our lives substantially! It will affect every aspect of our daily habits and possibly lead to greater changes than space travel, nuclear warfare, or the rise of the internet—at least on a personal, everyday life basis. It will lead to an increase of safety, comfort, and efficiency in many ways. The role of vehicles will change dramatically for all of us. On the other hand, we will finally face ethical, political, and technical challenges beyond today's imagination.

Let's take a look at the businessman—the traveling salesman for whom the vehicle is a professional tool to get in touch with customers and partners. Instead of wasting precious time behind the steering wheel of his vehicle and facing the risk of personal damage through accidents, the vehicle will take over the job. It will find its way on its own, and it will interact with the surrounding ecosystem and other vehicles in order to make the journey as comfortable and safe as possible. The traveling time is no longer lost—the businessman will be able to work in the vehicle and even meet and negotiate with business partners while being on the road. In a nutshell, the future is bright!

The perspective of the family man may be similar. His vehicle is a tool to get his kids to school, to go for daily shopping at the local markets, to drive to work and back home, and finally, to bridge long distances to spend the summer vacation in a remote location. An automated vehicle may take over the role of some non-existent or unaffordable private servant who is always at the service of his or her employer and takes the members of the family wherever it is required, but in a way, that is less dangerous, more efficient, and less costly. Finally, the kids will no longer face the dangers of a drunken or an inattentive driver. We believe that the family man will appreciate the benefits of the upcoming technology.

The driving enthusiast, on the other hand, may see things differently. What is the point of buying a $300K sports vehicle if the computer takes over control and operates the vehicle based on principles of being reasonable, cost efficient and "boring"? It defies the purpose! One may argue that computer control may not affect the world of a vehicle owner insisting on drive himself, but will he still be on the same road that is populated by vehicles with V2X networking and connectivity for which a distributed digital consciousness is dealing with all aspects of control and decision making? In the extreme case, the human being behind the steering wheel will be reduced to being an unreliable and unpredictable element in an otherwise "perfect" chain. The world of the driving enthusiast may no longer be the same once technology has taken over.

Critics of technology will obviously focus on the changes and potential dangers that will inevitably come with the introduction of automated and interconnected vehicles. Indeed, the number of open ethical and political questions is tremendous. Would we really want artificial intelligence to invade our lives at the anticipated

DOI 10.1515/9781501507243-001

extent? We may indeed place life-or-death decisions into the hands of a machine that may decide in microseconds who will be injured or who will die in an unavoidable accident situation. Will it be the driver, or the child who has unexpectedly run into the street? Will the computer make better decisions than the human being behind the steering wheel? In the end, at least, there is no one left to blame except the designer of the decision-making intelligence in the vehicle. All of us will still need to learn tough lessons in this space.

For technology entrepreneurs and corporations, finally, the business potential behind automated and connected driving is endless. It is certain that a move toward this technology will create massive employment for highly qualified professionals of information technology and data analytics.

The city is becoming a hostile place for the vehicle industry. Between traffic, environmental legislation, and parking challenges, many cities seem to have turned against the vehicle. Some heavily burdened municipalities worldwide have run out of alternatives to lower an excessive dust and nitrogen oxide load; for instance, Germany's top cities, Stuttgart and Munich have considered a citywide driving ban for diesel vehicles in their cities. How can we adapt to regain cities' favor, capture the interest of the urban consumer, and be a key part of the mobility mix? How do we embed mobility services like Uber, MyTaxi and Lyft that complement existing transportation options like the tram, railway, and bike?

Drivers and passengers increasingly presume access in their vehicles to connected services such as weather apps, transportation and travel information in the same way they can access these services with their smartphones. Real-time data delivery of web, radio, or video, and access to multimedia content, gaming, and social networks is already implemented or is frequently requested. As vehicle users continue to request remote services, usage data, and automated vehicle function support, the automotive industry has responded by investing heavily in a rapidly developing market of connected vehicles.

But we also perceive extremely divergent trends in how vehicles are used and viewed by their drivers and passengers. There is sometimes boredom for people who are riding in the vehicle, while the drivers and passengers may be extremely occupied and overwhelmed by complex situations, smartphone usage, and higher levels of automation inside and outside the vehicle. Both situations can very often become life threatening. These days, we are asked or ask ourselves why we want vehicles that drive themselves.

For most of us, learning to drive was a rite of passage to becoming an adult. This rite may soon change with networking vehicles and autonomous driving. The concept of not dealing with the rush-hour traffic does sound appealing, and whether we like it or not, autonomous vehicles are coming. Supported by computing and communications, autonomous vehicles offer opportunities to make our roads safer, improve fuel efficiency, and give us the added bonus of a more relaxing driving experience. That's how it looks today from the driver and passenger's seats.

The view from vehicle manufacturers, computing, communications, and other vehicle ecosystem-entering stakeholders might be quite different. For example, for computing stakeholders such as data center operators, the pervasive use of autonomous vehicles creates unprecedented challenges. We do see in the research, development, and test phases of autonomous vehicles that the amount of data coming from self-driving vehicles is huge. The autonomous vehicle, driven about 90 minutes a day, generates about 4 terabytes (TB) of data. Some vehicle trials are generating over 1 petabyte (PB) of data per month. By 2020, a single autonomous vehicle might produce 4 TB of data during 1 to 2 hours of driving. For comparison, a single drone flight captures up to 50 GB of data, and a fleet of 500 drones creating maps can record about 150 TB of data. Data centers are already one of the crucial building blocks for gathering, dealing, and analyzing that data for autonomous vehicle systems. The transport of vehicle data to the data center is a challenge in and of itself. To scale out support for, let's say, 10 million vehicles by 2020, data centers will have to deliver extraordinary performance in computing, networking, and storage.

Ultimately, to pave the way for the widespread use of autonomous vehicles, all vehicle ecosystem stakeholders need to be uniquely positioned. This is demonstrated, for example, by the many European research projects for vehicle networking, connectivity, and communications as part of European road transport research advisory council's "Vision 2050." Among these projects are CARTRE, SCOUT, COMPANION, AMIDST, AUTOPILOT, ENABLE-S3, AutoNet, and CarNet. The European road transport research advisory council calls *connectivity and networking* key challenges for autonomous driving, asking for solutions to balance the fast evolution of and request for connectivity with the slower pace of vehicle and infrastructure development, and to deal with the growing demand for communications, bandwidth, and data. The networking technologies, which span from vehicle to cloud, will require many distinct systems and capabilities to work together seamlessly to deliver tomorrow's self-driving vehicles. These are thrilling times for the vehicle ecosystem heading toward automated and autonomous driving solutions. But these times are by no means easy ones for the ecosystem stakeholders like vehicle manufacturers and their suppliers. It's a cutthroat challenge for any vehicle manufacturer or solutions provider today to determine which features the public will want, need, accept, and buy with the vehicle. Aggravating that challenge, it becomes even more demanding to estimate the types of advanced driving solutions that vehicle drivers and passengers are striving for—technical edge, safety, the joy of driving, or best comfort.

Obviously, the vision outlined in the examples above will not occur over night. Rather, a gradual, step-by-step introduction of the technology is going to be applied. We currently see first elements in today's high-end vehicles, such as driver assistance systems, (partly) autonomous driving within certain limitations, etc. These features will further evolve until vehicles will even be able to fully operate without any driver.

According to the U.S. Society of Automotive Engineers (SAE) and the German Association of the Automotive Industry (VDA), six levels with increasing degree of automation are defined for automated driving (SAE Standard J3016, January 2014) as illustrated in Figure 1.1. Starting from no automation (Level 0), there is driver assistance (Level 1), partial automation (Level 2), conditional automation (Level 3), high automation (Level 4), and finally, full automation (Level 5). At Levels 0 to 2, the human driver monitors the driving environment, while at Levels 3 to 5, an automated driving system monitors the driving environment. Increasing the level means raising the needs for communications both inside the vehicle and to the external world. The communications needs are strongly related to the degree of cooperation, such as co-operative sensing and manoeuvring. We use the SAE model throughout the book. Descriptions of the levels are included below in Figure 1.1 and Table 1.1.

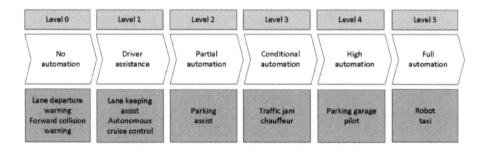

Figure 1.1: Automated driving levels according to SAE and VDA

Table 1.1: Levels of automated and autonomous vehicle driving according to Bundesanstalt für Straßenwesen (BASt), German

Type	Description of degree of automa-tion and the driver's anticipation	Example system
No automation	The driver guides the longitudinal (acceleration and deceleration) and the lateral (steering) guidance always during the entire drive.	There is no driver assistance system active in the longitudinal or transversal control.
Driver assistance	Drivers always perform either the transverse or longitudinal control. Individual other tasks are done within certain limits by the system. The driver must always monitor the system and must always be ready to take over the vehicle at all times.	Adaptive cruise control (ACC) with longitudinal control with adaptive distance and speed control park assist and lateral guidance by parking assistant.
Partial automation	The system implements transverse and longitudinal control for a certain period of time in specific situations. The driver must always	Motorway assistant with automatic longitudinal and transverse control on motorways up to an upper speed limit. The driver must

Type	Description of degree of automation and the driver's anticipation	Example system
	monitor the system and must always be ready to take over the vehicle at all times.	always monitor and react immediately to the request.
Conditional automation	The system gets activated and implements transverse and longitudinal control based on an optional takeover request. The driver receives additional data relevant to the automated function status. The driver must always monitor the system and must always be ready to take over the vehicle at all times.	Traffic jam function with automatic longitudinal and transverse control on motorways up to an upper speed limit. The driver gets the option to let the system to take over and must always monitor and react immediately to a request.
High automation	The system implements transverse and longitudinal control for a certain period of time in specific situations. The driver does not have to monitor the system always. If necessary, the driver is requested to take over with sufficient notice. System limits are all recognized by the system. The system is incapable of producing the risk-dominated state from any initial situation.	Highway driving with automatic longitudinal and lateral control. On motorways up to an upper speed limit. The driver does not have to monitor always and must react within a certain time reserve after a takeover request.
Full automation	The system performs completely transverse and longitudinal control in a defined use case. The driver does not have to monitor the system. Before leaving the use case, the system prompts the driver with sufficient time to pick up. If this is not done, the system is returned to the risk-minimized system state. The system limits are all recognized by the system and the system is capable of returning to the risk-critical system state in all situations.	Highway pilot with automatic longitudinal and transverse control on motorways up to an upper speed limit. The driver does not need to monitor. If the driver does not respond to a request for acceptance, the vehicle brakes to a standstill.

However, we believe that, aside from technology and business, there is a greater angle to the introduction of automated driving and networked vehicles that will be key—the gradual introduction of different parts of our road infrastructure. As a first step, we anticipate that autonomous driving will be made available on major highways linking the larger hubs of a given country. Such an environment indeed allows for a confined and reasonable level of deployment of roadside communications equipment and other required infrastructure, and is thus a reasonable business model. At the same time, the impact to the population will be considerable—once the driver has entered the highway, the artificial intelligence within the vehicle will take over, which is fed with information through a suitable wireless ecosystem. Long trips will become easier, and the problem of drivers falling asleep will no longer be relevant. Your vehicle will be able to take you all alone from Munich to Berlin, from Marseille to Paris, or from New York to San Francisco. The driver only needs to take care of the first miles to the highway, and the miles from the highway to the final destination.

Subsequently, the technology will be made available in areas outside of the major road infrastructure. First, major hubs such as large cities will be equipped, and perhaps eventually, smaller towns, villages, and countryside roads. It remains to be seen whether a full availability of automated driving will be available everywhere. While the deployment of the technology is straightforward along major roads, the full coverage of any residential area in any town and village may come at a cost-benefit level, which is unattractive—at least in the short- to mid-term. We will face a specific challenge during the deployment phase when some vehicles will rely on the new technology while others will not. We observe in particular that local drivers apply behavior on the road that is not written in a textbook or otherwise documented. An automated vehicle will need to adapt to such behavior. To provide an anecdote, one of the authors of this book recently experienced some local particularity in the beautiful south of Italy. On a two-lane road, some drivers honked their vehicle horns to forcibly create a third, imaginary middle lane. The vehicles on the left moved further to the left, the vehicles on the right moved further to the right, and a new middle lane was created, improving overall traffic efficiency. Who knows what other unwritten and undocumented ways of behaving on the road may exist across the world! It will be a clear challenge to the technology to fit into the existing ecosystem and to possibly adopt local habits. Therefore, we will further elaborate on the ethical, political, and technical challenges of the technology, and we will comment on the road ahead.

1.1 V2X objectives

There are wireless-based vehicle-to-vehicle (V2V) and vehicle-to-infrastructure (V2I) links (see Figure 1.2) to network or connect vehicles using local- and wide-area radio access network technologies. We use the term "networking" as an exchange of data between self-driving vehicle system components inside and outside of the vehicle. The term "connecting" refers to the setting up of communications links between these system components. The primary objectives behind V2V and V2I (referred to as vehicle-to-everything, or V2X) are to prevent accidents and save lives by alerting drivers to the hidden dangers that can't be sensed by on board equipment. Other names sometimes used for V2X are "cooperative connected vehicles," or "cooperative intelligent transportation systems" (CITS). By sharing data from any V2X-equipped vehicle within a specified radius, the driver can be alerted to the most common causes of accidents in time to take action. For example, for V2I, these include safety red light violations, curve speed warnings, weather alerts, and signals for work zone safety, bridge height, and pedestrian crossing. V2V safety applications and services are emergency brake light warnings, forward collision warnings, traffic alerts, emergency vehicle notifications, and road hazard detections. V2X convenience uses are eco-driving, parking information, truck platooning, speed harmonization, queue warning, and insurance pricing. And there's a lot more we need data services for— gathering behavioral data of the vehicles to improve them by applying machine learning, upgrading software and security features, or having vehicles exchange data with one another through wireless links.

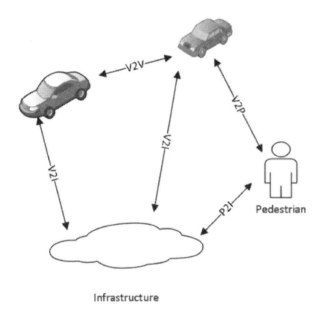

Figure 1.2: V2X networking and connectivity

V2X communications shouldn't be mixed up with autonomous and automated driving vehicles, which is a secondary target for this technology. As we can see from many tests of autonomous and automated vehicles, the capabilities of self-driving vehicles could be extended optionally by implementing communications networks both in the immediate vicinity (for collision avoidance, for instance) and far away (for congestion management, for example). But by networking vehicles with the traffic infrastructure, providing additional data for the internal data processing, one could no longer regard the vehicle's behavior or capabilities as autonomous. The vehicle is rather automated in this case. In other words, if vehicles are networked and cannot drive without the network, then they are not autonomous—they are purely remote controlled. A vehicle that couldn't drive when the access to the networks went down or went out of coverage is going to be unsafe and unusable.

In principle, self-driving vehicles are possible without networking vehicles with everything, even for high and full automation. There have been test pilots of self-driving vehicles based only on on-board sensor actor systems, like the Audi A7 550-mile piloted drive from Silicon Valley to Las Vegas in 2015 (Audi Press release). This pilot used high-resolution digital maps, which were downloaded offline from a server. V2X is specified to provide vehicles, drivers, passengers, and other stakeholders with additional data that will be integrated with data from the many other vehicles' on-board sensors and actors. V2X is an enabling technology that will make autonomous and automated driving vehicles much safer by helping them to extend their viewing and visibility range. Besides communications performance, reliability, security, and privacy are must-haves for V2X communications.

In this book, we start looking at the major ecosystem stakeholders for networked vehicles, and their basic usage scenarios and use cases identified so far, for specifying the requirements of research and development platforms being capable of up to Level 5 automated driving. As previously mentioned, Level 5 automated driving means, for example, running a fully automated robot taxi with an auto pilot, including self-navigation, collision avoidance, automated valet parking, highway chauffeuring, lane change management, steering, and throttle and brake actuation control. These research and development platform architectures consist of hardware architecture and system enabling software such as device drivers, computer vision libraries, mission and motion planning software, perception detection and localization software, and sensor and actuator software drivers. All these system components need to communicate internally and externally with each other reliably, safely, and securely via communications hardware and software. V2X communications has to provide the right performance for the various use cases to connect platforms comprised of components like cameras, LIDAR, and GNSS with computing and storage systems, for example, to accurately determine a vehicle's location. Already, today's advanced driver assistance systems (ADAS) like automatic cruise control (ACC), lane departure warning (LDW), and pedestrian detection (PD) form the backbone of tomorrow's mobility. Vehicles communicate with each other and with infrastructure. Vehicle-to-vehicle (V2V) communications allows vehicles to exchange relevant information such as local traffic data (for example, nearby accidents) and their driving intentions. Vehicle-to-infrastructure (V2I) communications is used to optimize the road network usage and thereby helps to reduce environmental pollution.

1.2 V2X and D2D networking and connectivity

The explosive growth in data traffic demand on wireless communications due to the popularity of networked and connected vehicles can be satisfied with several local- and wide-area network and access architectures. One of them is the exploitation of device-to-device communications (D2D) for connecting vehicles in close proximity within a common communications range. In D2D communications, vehicles communicate with each other without intermediate nodes. The inherent lower delay in communications that is required in some of the traffic safety use cases or for collision avoidance systems can be advantageous. By using D2D or V2V, respectively, rather than using V2I and relying on infrastructure, vehicle grouping—including group handover—can be done with minimal signalling from the network. Device-to-device communications has been introduced in 3GPP Release 12 as a side link communication, but only a few years later, the same standard has been optimized to vehicle communications requirements. Strict latency and communications requirements have been introduced. The D2D side link channel has been modified to meet those requirements and enable new use cases.

V2X communications is the transmission of data from a vehicle to any unit that may impact the vehicle, and vice versa. V2X communications consist of four types of communications: vehicle-to-vehicle (V2V), vehicle-to-infrastructure (V2I), vehicle-to-network (V2N), and vehicle-to-pedestrian (V2P). This data exchange can be used for safety, mobility, environmental, and convenience services and applications, including driving assistance, vehicle safety, speed adaptation and warning, emergency response, navigation, traffic operations and demand management, personal navigation, commercial fleet planning, and payment transactions. The basic component of a V2X system is the vehicle (V) and its connectivity to any other intelligent transportation system (ITS) component. V2X consist of transceivers located on vehicles, on the roadside infrastructure, in aftermarket devices, or within any other mobile devices like smartphones.

3GPP TR 22.885 defines roadside unit (RSU), vehicle-to-everything (V2X), vehicle-to-vehicle (V2V), vehicle-to-infrastructure (V2I), and vehicle-to-pedestrian (V2P). A RSU comprises an eNB or stationary UE transmitting or receiving data from another UE using a V2I application to support V2I services. V2X is a communications service where a UE transmits and receives V2V application data via 3GPP transport. V2I service is a specific V2X service between an UE and a RSU by means of a V2I application. V2I service is another specific V2X service between an UE and the communications network—in particular, a LTE network. V2P service is a specific V2X service between UEs using a V2P application and V2V service is the V2X service type where UEs communicate via a V2V application.

In this book, we use "networking vehicles with everything" as synonymous with V2X communications. Examples for V2X use cases are vehicle-to-vehicle (V2V) communications for platooning, convoying, and vehicle safety like collision avoidance;

pedestrian-to-infrastructure (P2I) for traffic light control and warnings; and vehicle-to-infrastructure (V2I) for smart intersection control and phased traffic light, dynamic environmental zones, and real-time vehicle localization. Vehicle-to-everything (V2X) is any communications involving a vehicle as a source or destination of a message. Depending on the nature of the other communications endpoint, several special cases exist: vehicle-to-vehicle (V2V), vehicle-to-infrastructure (V2I) (road infrastructure, which may or may not be co-located with cellular infrastructure), vehicle-to-network (V2N) (e.g., a backend or the internet), vehicle-to-pedestrian (V2P), and so on.

1.3 Technical challenges for V2X

Networking and connecting vehicles internally and externally have been inspiring research and development topics for decades. The tremendous progress in computing, communications and sensors actors could enable fully networked vehicles and in particular automated and autonomous vehicles starting in big quantities in 2020. Sensor actor technology, massive and robust signal and data computing, and enormous advances in wireline and wireless communications technologies will make networking and connecting vehicles with everything happen. The autonomous and automated vehicle platform is based upon sensors and actors, computing, storage, communications, and software. We derive some major technical performance indicators of networking and connecting vehicles, taking into account the data sheets of state-of-the-art vehicle platform components and most challenging use cases.

The way forward to autonomous and automated driving with increasing automation starts from data sharing and raising, and ends with decision sharing. At level 1, the vehicle gains awareness with status data like, "I'm a vehicle" and responsiveness regarding its location and traveling direction. At level 2, sensor data like, "It's raining here" or "I just passed a RSU" are added. At level 3, the vehicle acts cooperatively, requesting the driver to confirm data and system actions in use cases like lane merging or traffic jam assist. At level 4, the vehicles work in collaborative or cooperative ways in use cases like "Let's platoon" or slot-based intersections. Finally, at level 5, the fully automated vehicle relies on massive sharing and fusion of sensor data and distributed computing.

1.3.1 Sensors

The up-to-date technology for highly autonomous and automated vehicle driving in controlled and in real environments is quite advanced. Vehicles use state-of-the-art sensors like radar, LIDAR, ultrasound, GNSS, and video camera systems (Figure 1.3) jointly with high definition (HD) maps, which enable on board computing platforms to identify obstacles, relevant traffic signage, and appropriate navigation paths.

Long-range radar is taken for adaptive cruise control (ACC), forward collision warning, and night vision, and LIDAR for object detection like pedestrians and collision avoidance. Video cameras are deployed for lane departure warning, traffic sign recognition, surround view and park assist, short- and medium-range radar for forward collision warning, pre-saving, back-up obstacle detection, stop-and-go, low speed ACC, cross traffic warning, blind spot detection, and rear collision warning. Ultrasound is sometimes used for park assist.

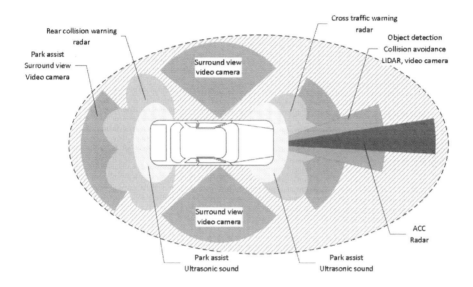

Figure 1.3: Enabling technologies in a vehicle—sensors

We view V2X communications as an enabler to enhance any sensor in the vehicle. While sensors like radar, LIDAR and video cameras are actively surveying the environment around the vehicle, V2X wireless communications with non-line-of-sight capability, is able to extend its range far beyond, to any range needed by communicating with other vehicles, infrastructures, and pedestrians. We look as well at other in-vehicle sensors, which play a major role to ensure the main purpose of the vehicle—to drive. Position sensors in the vehicle are used, for instance, to measure the angles of the engine throttle plate, the chassis height link bar, the fuel level via float arm, and the steering wheel angle. Vehicle pressure sensors are used to determine the pressure of the engine manifold absolute, ambient barometric, evaporative fuel system leak, brake fluid, chassis adaptive hydraulic suspension, air conditioner compressor, or the common-rail fuel injection. Temperature sensors are there for engine coolant, fuel, brake, and steering fluid levels. Mass airflow sensors measure steady state and transient mass flow of air into a vehicles' engine. Torque sensors estimate the steering wheel torque for electric power steering (EPS), driveshaft (transmission-out) torque, and clutch shaft (engine-out) torque. Linear acceleration inertial sensors measure the vehicle stability and chassis adaptive suspension systems, vehicle frontal, side, and rollover crash sensing and engine knock detection. Angular rate (Gyro) inertial sensors are there for vehicle electronic stability control (ESC), active chassis suspension, rollover protection of side curtain airbags, and vehicle navigation systems. Chemical and gas composition sensors measure gas, oxygen, and oil characteristics. Comfort and convenience sensors estimate solar radiation, twilight, dimming of mirrors, fluid levels, rain detection, temperature, and noise. Driver, passenger, and security sensors are there to determine occupants' weight, size, position, seated weight, seatbelt tension, buckle status, seat position, and intrusion detection.

Nowadays, in vehicle tests, long-distance radar, LIDAR, short-range radar, and video data get uploaded to servers and clouds for analysis. In commercial use, this enables the application of deep learning and KI for vehicles in an ongoing process now. The other above-mentioned sensor data might additionally to get stored offline for big data analytics and telemetry as a service.

1.3.2 Computing

A vehicle today is a distributed computing system with—when all optional equipment is installed—50 to 150 electronic control units (ECU) and distributed storage up to several GBs which run vehicle control, navigation and telematics and infotainment and accomplish vehicle performance and behavior. ECUs are connected with several in-vehicle networks with high real-time reliability and safety requirements. There is an ongoing transformation of the computing architecture from distributed ECUs toward a more centralized and domain-oriented electronic control (CEC) system to reduce the large number of ECUs and, in particular, to save space in order to support autonomous and automated driving. The storage requirements are changing, as well, and are driven by HMI, IVI, and data recording.

Computer vision, cognitive computing, deep learning, and AI for autonomous and automated driving require massive on-board computing power for real-time decision making, since vehicles have to make critical decisions in real time independently from any cloud. Nevertheless, cloud-computing is also required for collective and deep learning data acquisition in vehicle tests as well as common use—therefore, vehicles are becoming part of a wider networked and connected ecosystem. Use cases are ADAS sensors, big data, and analytics applications in which vehicle sensors are used to create, maintain, and improve a data source for services and applications (for example, dynamic HD maps). Vehicles—in particular, next generation autonomous vehicles—effectively become mobile data centers. The vehicles themselves will generate and process massive amounts of data from on board sensors, but will also take in large quantities of data from the network, including ultra-high definition maps and near real-time information to help navigate and detect what's coming around the next corner or to avoid upcoming traffic congestion.

Vehicle computing platforms evolve with telematics, transmission control units (TCU). We identify three major architecture concepts delivering sufficient processing power for autonomous and automated driving, telematics and IVI which are getting prepared for V2X communications as well.

1.3.3 Networking

Today's cellular communications systems are not designed to handle the massive capacity required to support millions of self-driving cars when they hit the road. The way toward autonomous and automated driving involves the evolution of current communications technologies, as well as the development of new ones such as 5G. Networking and connectivity are important components of the upcoming autonomous and automated vehicles ecosystem where secure, reliable V2X communications with low latency is mandatory. V2X technologies involve the use of several wireless technologies to achieve real-time two-way communications among vehicles (V2V), between vehicles and infrastructure (V2I), and pedestrians and vehicles (PDI). The convergence of sensors, computing, storage, and V2X networking and connectivity will stimulate autonomous and automated driving.

V2X networking and connectivity can be supported through different wireless technologies. V2V networking and connectivity is likely by means of dedicated short-range communications (DSRC) technology, developed with safety critical use cases in mind and based on the IEEE 802.11p Wi-Fi standard. IST-G5 and WAVE as of Table 1.2, specify dedicated short range communications (DSRC) for an ad-hoc 2way network in dedicated licensed spectrum bands with 7 channels. DSRC has got low latency below 50 milliseconds and a range up to 2 kilometers with data rates from 6 to 27 Mbps. It provides signed messages using a public key infrastructure (PKI).

Table 1.2: V2X networking and connectivity stacks as of IST-G5 and WAVE protocol stack

	IST-G5 (Europe)	WAVE (USA)
Security IEEE 1609.2		
	C2C-CC	SAE J2945
	CAM, DENM, SPAT, MAP	SAE J2735 (BSM, SPAT, MAP, etc.)
	GeoNet / Decentralized congestion control	DSRC WAVE short message protocol
	TCP/IPv6	TCP/IPv6
	IEEE 802.11p MAC EU Profile	IEEE 802.11p MAC DSRC
	IEEE 802.11p PHY EU Profile	IEEE 802.11p PHY DSRC

This is the path the United States' National Highway Traffic Safety Administration (NHTSA) and other regulators and stakeholders are currently assessing. For instance, General Motors the largest vehicle manufacturer in the U.S. said in March 2017, that all of its new 2017 Cadillac CTS vehicles would come armed with DSRC based V2V networking and connectivity. And Volkswagen in Germany said in June 2017 that it will fit first VW models with pWLAN technology which is based on IEEE 802.11p in 2019.

At present, the vehicle industry's safety stakeholders generally are not fully committed to whether cellular technologies can be used besides infotainment for safety critical ADAS and autonomous and automated driving functions. This is due to performance indicators like cellular network coverage, reliability and liability, security and the need for mobile operator subscriptions, and roaming agreements. Issues like these compete unfavorably against—LTE with DSRC technology, for instance—in avoiding collisions and other safety applications as of today.

That's where next generation new radio communications technology comes in. It's expected to deliver more capacity, ultra-low latency, faster speeds and vehicle-to-vehicle (V2V) connectivity for the era of autonomous vehicles. With the increase in mobile traffic created by making everything smart and connected—including vehicles—networks need to transform to handle the additional connections. By leveraging innovations from cloud data centers, the network infrastructure is becoming agile, software-defined and flexible to enable the efficient and ultra-low latency connections needed for autonomous driving. Autonomous vehicles will need to act on near-instantaneous updates from around the corner or down the road—there's no time to send or receive data from server hundreds of kilometers away when your vehicle is platooning with 100 other vehicles at 100 km/h down the highway. With 5G, networks will deploy computing resources at the very edge of the network in cellular base stations and towers that will deliver road status updates to connected vehicles in milliseconds.

With the millimeter wave spectrum and advances in wireless and antenna technology, 5G is expected to deliver multi-gigabit speeds for mobile uses. The industry expects 5G speeds to eventually be capable of up to 10 gigabits per second, which is over 600 times faster than today's fastest average LTE speeds in the U.S. Self-driving vehicles will opportunistically connect to 5G cells when available and needed, then seamlessly fall back to 4G LTE to maintain network connectivity. 5G aims to deliver multiple models of connectivity—including direct vehicle-to-vehicle connections, as well as vehicle-to-infrastructure or vehicle-to-network connectivity. This flexibility in 5G design anticipates the varied situations that self-driving vehicles will encounter in, around, and between cities.

We show that there are technical challenges on top of the communications link for networked vehicles in the dynamic system architecture through the reconfiguration of a vehicle communications system after delivery, the reuse of system and software components, and building blocks during development, tests, and trials. There are also challenges in the operation of vehicles and the traceability of requirements and proof of their realization in variable and dynamic systems.

If we consider necessary updates of hardware and software features for networked vehicles, harmful influences on other system components must be absolutely excluded. This applies not only to direct functional relationships but also particularly to the resource requirements and the real-time behavior of the networked vehicle system, as well as aspects that could compromise the functional safety. Particularly high requirements arise when functions of different criticality, such as in-vehicle infotainment functions and vehicle functions, are executed on the same hardware. For this purpose, technologies are required that allow close interlinking and sharing of resources between the mentioned system parts, but do not endanger the functional safety of the vehicle functions.

Certain use cases of V2X are very challenging when it comes to vehicle functional safety according to ISO 26262, which requires that if and when a vehicle misbehaves, it is not hazardous to any life. Vehicle networking and connectivity are going to require additional hardware and software features in terms of modular hardware and software (with embedded security, in particular) for autonomous and automated driving. In order to ensure that these features fulfil the ISO 26262 requirements, a common validation and certification procedure is a prerequisite—the communications between all involved vehicles, infrastructures, and pedestrian components and functions is required to be evaluated how V2X networking and connectivity reacts in the case of misbehavior.

We need to ensure the best use of V2X networking and connectivity, so it is necessary to take into account the potential system impacts which cannot be anticipated a priori for all use cases. This requires new solutions that allow monitoring, plausibility checks, and safety-related interventions during vehicle operation via networking. These solutions must ensure compliance with clearly defined safety criteria across functional and system boundaries. In addition, appropriate services and interfaces will be required to enable early identification of anomalies, which may require the permanent collection and centralized evaluation of diagnostic data in real-time.

The introduction of new and open interfaces to the networked vehicle, and the possibility of reloading and adapting software during runtime, create new threat scenarios regarding data security. Therefore, appropriate precautionary measures must be taken in terms of processes, architectures, protocols, but also services at hardware and software levels. In the short-term, this affects the unintended accesses to the non-volatile memory. This can be solved with currently available evolved security mechanisms such as authorization of flash memory, detection of software manipulation, and the transmission of successful security patterns from computing and communications like firewalls or virus scanners. In the mid-term, the focus shall be on the threats at run-time, particularly through the communications interfaces created by networking and the reloading of software. There are proposals to use security standards (such as IEC 62443) with their specified automotive safety integrity levels (ASIL) for compliance to safety-critical V2X components.

An increasingly important aspect is the confidentiality of private data (privacy). For driver and passenger protection, basic services are required, which are exemplarily secure communications channels, efficient encryption, reliable anonymization, and pseudo-authentication. The privacy services shall be adapted to altered threat scenarios—for example, by updating cryptographic procedures. With regard to the prevention or detection of manipulations via communications interfaces or by reloaded software, services are required for the certification of software, the reliable authentication of accessing third parties possibly in conjunction with a trust center, intrusion detection, and for the quarantine and isolation of suspicious software units.

For more than ten years, cellular networking and connectivity in vehicles was seen in a first phase as a solution to calling for roadside or emergency assistance, as in the case of General Motors' OnStar or Daimler's mbrace® service. These services are still vital and are part of a broader set of upcoming safety services and applications. But in the last few years, there has been a second phase of vehicle networking that is driven by smartphones and vehicle infotainment. Applications like Google Maps, Twitter, and Facebook were integrated into in-vehicle infotainment (IVI) systems along with weather, video, and music streaming applications. Audi integrated vehicle-to-cloud-to-infrastructure (V2C2I) networking and connectivity to its Audi connect infotainment systems, enabling traffic light signal phase and timing (SPAT) application (currently only working in Las Vegas in the U.S.), which informs drivers of the phase status of an approaching traffic light. It went further by implementing Wi-Fi, Bluetooth, and NFC-enabled remote access to vehicle data, diagnostics features, door lock, and even remote engine start. The most current advances are speech recognition, parking reservations, and digital assistants delivered via connected smartphones. BMW, Daimler, and Volvo introduced vehicle-to-cloud-to-vehicle (V2C2V) networking and connectivity to enable communications between their vehicles in 2017. On the opposite end, neither Alphabet (Google) nor Tesla Motors—frontrunners in automated driving—are currently leveraging wireless cellular or Wi-Fi technology for collision avoidance and automation.

BMW's roadmap for deploying autonomous technology shows achieved SAE level 2 advanced driver assistance that still requires hands-on direction from a human driver at all times. Level 3 automation will be achieved with its iNext vehicle by 2021. BMW's iNext is going to be technically capable of Level 4 or 5 autonomy by 2021 for development and tests, but not for production. BMW uses high-definition maps to extend vehicles' reference data further than the range of sensors. Together with V2I communications, this allows a vehicle to navigate itself to a stop in a safe location if the driver is asleep or otherwise unable to immediately take control. Artificial intelligence is seen as another key technology for safe self-driving vehicles, but will need to overcome significant challenges associated with AI data processing and communications. BMW expects fully self-driving vehicles between 2020 and 2030.

When 5G technology approaches, it will transform wireless vehicle networking and connectivity. This transformation of wireless networking and connectivity will come to play a decisive role in many vehicle domains, including autonomous and automated driving use cases. It's becoming evident that the vehicle is increasingly dependent on tight integration between networking and connectivity, sensors, actors, storage, and computing. V2X will be enhanced by the capabilities provided by 5G as another option to DSRC for low latency collision avoidance, V2V, V2P, and V2 networking and connectivity. The challenge is to make cellular networking and connectivity a safety and mission-critical technology with the prospect that general wireless technology (cellular, in particular) is becoming an essential part of autonomous and automated driving vehicle solutions.

1.4 Society, ethics and politics

We are clear on one thing—the vehicle is the most favored toy. Many of us are distrustful of leaving our personal rights and control over the road to the vehicle. For us, the vehicle is not just a means of transportation from A to B. No—our vehicle means, among other things, passion, status, freedom, fun, frustration, or even individuality. Who doesn't enjoy being able to speed up his or her own style and be the master of the situation? Will this be taken away from us with a networked, autonomous vehicle? If it is only a matter of pure movement, public transportation is available to most of us. So, how will the world work with automated vehicles in the future? Can an autonomous vehicle honk, even if there is no immediate danger? And how will a vehicle convoy on Munich's Leopoldstraße look like on the occasion of winning the Football

World Cup in 2034? True, the driver has both hands free to swing the flags. But a driving speed of exactly 50 km / h is likely to cloud the festive mood. Or is a vehicle convoy with step speed, a hazard warning system, and horns planned in the common mode? What about the times when a manually controlled vehicle with a supposed traffic-rowdy person at the wheel winds through the traffic jam, while the autonomous vehicles avoid and circumvent a potential accident? Through networking and connectivity, will all vehicles get informed accordingly and form an alley-free ride for the driver? Driving schoolchildren in Germany are allowed to park at the push of a button. According to the driving license regulations, all equipment and systems available are basically approved. Driving with a modern "computer on four wheels" is not comparable to a vehicle without any comfort. The driver will quickly notice the progress when they enter their own car. The question of what has enabled the development on the vehicle side is answered quickly: software (in conjunction with sensors and actuators) and the wiring system. These two "actors" of a vehicle lie concealed behind covers or the carpet of a vehicle and would have actually earned a more prominent place.

An INRIX study from May 2017 shows that for 56 percent of German drivers, the integrated technology is an important factor that influences the purchasing decision as much as the vehicle performance. According to a representative survey conducted by Bosch, only 33 percent of Germans show interest in autonomous driving. They fear, for example, loss of control or even hacker attacks. In China, the survey is an astonishing 74 percent. But where does this big difference come from? Does the autonomous driving fit our lifestyles and us? Do we perhaps underestimate the safety aspect of road traffic, which is driven by autonomous and automated vehicles?

How do the Germans perceive the driverless car? Results of the study on the autonomous driving of TÜV Rheinland in May 2017 demonstrate this. A total of 1,400 drivers were surveyed across all age groups. The representative online survey is surprisingly positive, but the scepticism increases with increasing age. Since the agreement of the Bundestag and the Federal Council it has been clear, the autonomous vehicles will come—it is only a matter of time. The automotive groups and suppliers are eager to develop products and systems. The number of today's driver assistance systems is growing steadily, and more and more test vehicles are in use around the world. But are the Germans also willing to use an autonomous vehicle and let themselves be driven? According to the first survey results, the German vehicle drivers were not yet convinced. The Figure 1.4 below shows survey results regarding the area of application. According to this survey, an average of 76 percent of the drivers interviewed can imagine a computer as a driver or are ready to use a driverless vehicle. A more detailed view of the supporters shows, however, clear differences, depending on the driver and use, as seen below in the Figure 1.5.

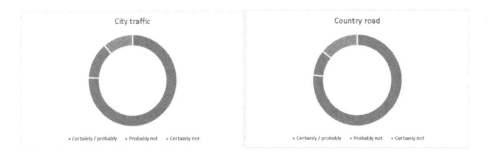

Figure 1.4: Acceptance of autonomous driving among Germany's drivers in 2017 by area (TÜV Rheinland.in May 2017)

Figure 1.5: Acceptance of autonomous driving among Germany's drivers in 2017 (TÜV Rheinland.in May 2017)

The rejection of autonomous driving appears to be the least where the technical challenges are highest: in urban traffic. It is striking in the analysis that, in particular, the younger generation of technology is very open-minded. Scepticism in technology increases with age. The seniors would, however, benefit from autonomous driving. There would be no need to ask for a new driving license or for mobility in old age. However, many people also see problems that are associated with autonomous driving. More than two-thirds are concerned about the difficulty of clarifying the debt issue and the liability in case of accidents. The initially important question about the alternatives in the case of unavoidable accidents has moved backward in prioritization. More concerned are the interviewees that hackers or cyber criminals supposedly manipulate the independent car. In addition to the above-mentioned questions about the basic readiness for autonomous driving, the focus of the study was also on the importance of system auditing and data protection, which around 90 percent of drivers see as important.

The automotive industry is racing to develop vehicles that are more and more autonomous, but the road ahead is tricky to navigate. Drivers are used to being in control. If the stakeholders push too hard and too fast, drivers won't go along for the ride. It's very much about the agency and control in the cockpit and how drivers and passengers want to interact with their connected vehicles. Today, drivers trust their vehicles because they themselves are in control. We can imagine that vehicles support some of our needs and give us choices. We can aspire that vehicles could predict our needs and make decisions for us. But the more our vehicles take control, the less we, as drivers, trust them. We don't want to get overwhelmed by too much data and urgency. The vehicle has to be highly personalized and adaptive, and has to intelligently determine the right options to the user at the right time. But all in all, this is a very treacherous path that may eventually lead to a reduction in personal freedom and trust, as a certain number of drivers and passengers feel today.

Networked vehicles and autonomous driving are going to be the next big disruptive innovation in the years to come. We consider it as being predominantly technology-driven and therefore, we suppose it to have massive societal effects in many areas. When we radically reconstruct the transportation and travel infrastructure by networking vehicles with everything, we are also going to alter how our municipalities and neighborhoods look. We are going to transform the population settling in our rural and urban areas, and our economy, society, and culture.

A Nuance and DFKI study (Nuance, DFKI: Cognitive and Conversational AI for Autonomous Driving May 2017) about the needs of intelligent and collaborative assistants to effectively engage passengers in networked self-driving vehicles shows that the top five activities in a networked autonomous vehicle would be listening to the radio (64%), relaxing (63%), talking on the phone (42%), browsing the internet (42%), and messaging (36%). If driving with passengers, the percentages would change—drivers and passengers would be naturally having more conversations (71%) or listening to the radio (58%), rather than talking on the phone (only 19%) or messaging (23%). Integrated, multimodal user interfaces leveraging voice, touch, and visual cues are considered more agreeable and effective than visual cues, leading to faster reactions than simply vibrations or haptic alerts. Drivers trust audible and haptic responses from the automotive assistant more than visual cues alone.

The interaction of human and machine casts new ethical questions in the time of digitization and self-learning systems. The autonomous driving of vehicles could even be requested if it is thereby possible to reduce the number of accidents to zero. But in highly automated road traffic, dilemma situations cannot be completely ruled out. The approval of automated driving systems also depends on the considerations regarding human dignity, personal freedom of decision, and data autonomy. To deal with these issues, for example, the ethics commission of Germany's Federal Ministry for Economic Affairs and Energy published guidelines (Bundesministerium für Verkehr und digitale Infrastruktur. Ethik-Kommission Automatisiertes und vernetztes Fahren. June 2017) for partially and fully automated driving systems for the improvement of the safety of all involved in road traffic in June 2017.

These guidelines concluded that the protection of human beings has priority over all considerations of usefulness. The responsibility for the introduction and approval of the networked driving vehicles remains with the state and under official control. In the case of legal structuring, the individual's right to free development must be taken into account. In dangerous situations, the protection of human life must always have priority over the prevention of damage to property and animals. In case of unavoidable inaccuracies, people must never be qualified according to personal characteristics such as age, gender, or physical or mental constitution. It is not software that decides, but the producer or operator of the autonomous vehicle driving system. Many dilemma situations in which the question is to live or to die can neither be standardized nor programmed. In any driving situation, it must be clearly regulated and recognizable who is responsible for the driving task—the human being, or the computer.

The report supplements the legislation for networked and automated driving launched in June 2017. Now the computer can partly take over the vehicle control. The last responsibility remains with the vehicle driver. The driver is obliged to resume the driving responsibility without delay if the highly or fully automated system prompts them to do so, or if they recognize, owing to obvious circumstances, that the requirements for the intended use of the highly or fully automated driving functions no longer exist.

Let's assume that more than 75 million vehicles are sold annually, and the number of vehicles is around 1.2 billion. Then it takes around 15 years for the whole vehicle armada to turn over. For example, if vehicle makers start producing nothing but fully autonomous cars in 2020, we will still see a mix of manual and semi-autonomous vehicles until 2035 or later. The growing vehicle production rates per year and the popularity of used vehicle marketplaces make this turnover rate even much longer. Consequently, regulation authorities, insurance, and politics will have had to catch up long before this point in time. Safety and security standards have to be put in place, and the question of liability in cases of accidents has to be answered. After the turn over, municipalities will look drastically different. Sidewalks and bicycle routes could go away, as pedestrians, bicycles, and vehicles share the roads. There could be no street parking, since parking areas could be dynamically allowed everywhere. We could also get rid of traffic signs and infrastructure, which could be exchanged with computing and communications devices that only need to communicate with vehicles.

There are major legal and policy challenges surrounding the networked autonomous vehicle ecosystem. For example, we see an urgent call for action in regulation and standard setting for critical event control, driver responsibility, ownership and maintenance (Teare, 2014), civil and criminal liability, corporate manslaughter (Browning, 2014), insurance, data protection, and privacy issues (Khan, et al., 2012). Regulation, certification, and testing are required in homologation, periodical main examination, and data protection of the collected data. Regular monitoring of data protection and ensuring the reliability of autonomous vehicles will further increase the acceptance of potential users. Networked autonomous driving vehicles require an adaptation of existing legal frameworks needed for traffic and vehicle regulations. The adapted regulations accelerate mass production and the commercialization of networked autonomous vehicles. So, we perceive many development plans and initiatives of worldwide public authorities with the current objective to pave the way for a step-wise introduction of networked automated vehicles. Many of these are built around standardization, regulation, testing, safety, or networked autonomous vehicle technology developments.

These activities from various stakeholders, including governments in Asia, Europe, and the U.S., are supporting or even advocating vehicle communications. The U.S., which is ahead of other countries in developing and regulating driverless vehicles, plans to set its specific rules (Bryant Walker Smith: Automated Vehicles Are Probably Legal in the United States). In the U.S., several federal states have already passed laws authorizing networked autonomous vehicles testing on their roads (Walker S., 2014). Legislatures in California, Nevada, Michigan, Florida, and Tennessee have passed bills enabling automated driving (CIS Automated Driving: Legislative and Regulatory Action). The National Highway Traffic Safety Administration in the United States provides an official self-driving vehicle classification very similar to SAE levels. Instead of SAE 6 levels it differentiates between the levels: no automation, function-specific automation, combined function automation, limited self-driving automation, and full self-driving automation.

China is one of the most ambitious areas when it comes to networking vehicles with everything. The HD maps have issued a roadmap for having highway-ready, self-driving vehicles in 2021 and autonomous vehicles for urban driving by 2030 (Li Keqiang, Tsinghua University). China may utilize wireless data communications technology (like LTE or 5G) that is already used in many vehicles to access the internet, and adopt it for vehicle-to-vehicle communications rather than the dedicated short-range communications (DSRC) standards developed in the U.S. and Europe.

Europe started adapting the Vienna Convention on Road Traffic and the Geneva Convention on Road Traffic (Reuters, 2014) in order to be able to allow networked autonomous vehicles, but legal issues and uncertainty are still there. Japan and major European countries cooperate to urge the U.S. to adopt common standards compiled by a United Nations expert panel as part of the World Forum for Harmonization of Vehicle Regulations. The regulations will contain principles, such as safety provisions, controlling autonomous passing to highways, and holding human drivers accountable for any accidents.

In March 2017, Germany announced that the German Road Traffic Law (StVG) will get complemented in that vehicles with automated systems (highly automated or fully automated) will be allowed to be used in traffic on public roads as such, and that the vehicle driver is allowed to hand over the vehicle control system to the technical system in certain situations. The autonomous driving system can be manually overridden or deactivated at any time by the driver, who is obliged to take over the vehicle control immediately when asked by the system.

We still perceive societal, ethical, and political issues that must be answered as we move closer to commercial services of networked autonomous vehicles. How should engineers build ethical choices into automated features and advanced computing and communications algorithms? What are the factors in the algorithms, if there are any that would lead its system to turn one way or another? There are a wide variety of ethical issues that remain, and system architects have to make choices regarding how to deal with them. Learning how to treat these complicated societal and ethical issues is a major challenge facing the way forward (J. F. Bonnefon, A. Shariff, I. Rahwan, "The Social Dilemma of Autonomous Vehicles"). In this book, we do not elaborate further on this important topic, because we think its importance requires a book of its own.

1.5 Outline

We treat in Chapter 2, the plethora of networking vehicles applications and use cases and provide insight into the many currently somehow disarranged views of computing, communications and vehicle stakeholders. We look at how large computing and communications stakeholders like Intel, Qualcomm and others define and specify vehicle networking and connectivity scenarios and use cases thereof. We reveal the view of big vehicle manufacturers like Audi, BWM, GM, Toyota and Volkswagen. And we look finally at the current hype in autonomous and automated vehicle networking and connectivity which is somehow climaxing in multi-stakeholder organizations like 5GAA, 5G-PPP, 3GPP, SAE, ETSI, GSMA and NGN. We provide important comments and offer the main use cases and scenarios from our view.

In Chapter 3 we show communications requirements for vehicle networking and connectivity as seen at this time by communications, computing and vehicle standards and regulation developing organizations mentioned in the previous chapter. Every camp has noticeably its own assessment and understanding of what kind of communications technologies are preferably to be implemented. The regulation of the spectrum resources in terms of frequency bands is another topic which needs to be solved soon, taking the development cycles of the vehicle industry and the time schedules of regulation bodies into account. We describe the challenges of dealing with these requirements to fulfill the future of autonomous and automated vehicle usage scenarios and propose a way forward.

We start with the state-of-the art technologies for vehicle networking and connectivity in Chapter 4 and look how the main vehicle platform components, sensors and actors, computing and communications building blocks are going to change for autonomous and automated vehicles. In particular we show the reader how these changes impact the communications technology opportunities which are currently available, together with the research and development challenges necessary to make cellular networking and connectivity a safety and mission-critical technology. LTE-A evolution and 5G cellular systems have the potential of supporting challenging and upcoming use cases that require low-latency, high reliability or high safety. Cellular V2X is able to work together with DSRC communications to enhance V2X communications.

At first glance, autonomous and automated driving does not need V2X communications. We take a second look and explain why V2X communications is just adding another sensor to the vehicle to support improved situational awareness, provide redundancy and make other sensors more reliable. V2X communications offer long range, data for collaborative driving and non-line-of sight capabilities. Based on the relationship of vehicle dynamics and communications requirements, we derive data throughputs for specific sensors and applications and map it against the performance of wireless communications technologies. We show that the ratio between uplink and down-link is symmetrical. V2X technologies seek to address, aside from automated driving and advanced driver assistance systems (ADAS), situational awareness, mobility services, and convenience services. There is the evolution of current ADAS, enabled by cameras and sensors, toward automated and autonomous driving where vehicle networking and connectivity augment ADAS and support it.

We elaborate in Chapter 5 on D2D and 3GPP release 12 side link communications and how it addresses the low-latency, high-reliability V2X use cases, for example in a complementary manner to DSRC. 3GPP LTE-based V2V achieves substantial link budget gains due to frequency division multiplexing (FDM) and longer transmission times. Hybrid automatic repeat request (HARQ) retransmissions are an option to realize higher link budgets. Turbo coding and single-carrier FDM implementation increase further link budgets. Combining sensing of the radio resources with semi-persistent transmissions at the system level exploits the periodic nature of V2V traffic and does not cause carrier sensing overhead. It results in better spatial reuse of radio resources.

We highlight in Chapter 6 the most widely implemented convenience service which is infotainment. Infotainment solutions for example in relationship with smartphones and software apps or the Genivi platform already provides today plentiful networking and connectivity solutions for V2X in many vehicles. Vehicles are integrating Android Auto, Apple CarPlay, Baidu CarLife or MirrorLink into their infotainment system. We look, in particular, at solutions to stream data into the vehicle and to upgrade and extend platform features via software using V2X and how these solutions get integrated with telematics, safety and connectivity planning. The current infotainment systems evolve into connected services, digital commerce, vehicle diagnostics, predictive maintenance, vehicle tracking and insurance.

We provide answers in Chapter 7 regarding the introduction of dynamically reconfigurable systems in networked vehicles which requires urgently an evolution of the existing framework conditions. In particular, the use of external software components for the dynamic expansion of a networked vehicle system poses challenges to the proven protection of these components. This needs to be done on the basis of a comprehensive component and building blocks specification, regulation and testing and, if necessary, must be assured by the certificate of a trustable stakeholder.

Corresponding contractual frameworks between the suppliers of external components and building blocks, manufacturers and owner, driver or passenger of a vehicle can accompany the use of the components and building blocks from copyright and other legal aspects. We use wireless communications examples to illustrate the offered principles and solutions.

We give our assessment on the current status of vehicle networking and connectivity and the evolution of local- and wide-area communications technologies in Chapter 8. With a wide-ranging view on major vehicle ecosystem stakeholders, we explain why we think that V2X communications are a critical component of the networked and connected vehicle of the future. We motivate why the wireless communications ecosystem stakeholders shall engage in early efforts to assess the approaching capacity and coverage needs of networked connected vehicles now in a tight collaboration with vehicle industries.

References

2020 Roadmap, European New Car Assessment Programme (EURO NCAP) June 2014.

3GPP TR 22.885 Study on LTE support for Vehicle-to-Everything (V2X) services. V14.0.0 21.12.2015.

5G Americas V2X Cellular Solutions. October 2017.

Adriano Alessandrini, Andrea Campagna, Paolo Delle Site, Francesco Filippi, Luca Persia, Automated Vehicles and the Rethinking of Mobility and Cities, In *Transportation Research Procedia* (2015), Vol. 5, pp. 145–160.

Audi: Mission accomplished: Audi A7 piloted driving car completes 550-mile automated test drive. Press release 04.01.2017.

Autoware: Open-source software for urban autonomous driving. https://github.com/CPFL/Autoware.

Bryant Walker Smith, Automated Vehicles Are Probably Legal in the United States, *1 Tex. A&M L.* (2014) Rev. 411.

Brandon Schoettle and Michael Sivak, A Survey of Public Opinion about Autonomous and Self-Driving Vehicles in the U.S., the U.K., and Australia. UMTRI-2014-21.

Bundesministerium für Verkehr und digitale Infrastruktur. Ethik-Kommission Automatisiertes und vernetzes Fahren. Juni2017.

CIS Automated Driving: Legislative and Regulatory Action https://cyberlaw.stanford.edu/wiki/index.php/Automated_Driving:_Legislative_and_Regulatory_Action#State_Regulations

ETSI TR 102 638 V1.1.1, Intelligent Transport Systems (ITS); Vehicular Communications; Basic Set of Applications; Definitions.

ETSI TS 102 637-1 V1.1.1, Intelligent Transport Systems (ITS); Vehicular Communications; Basic Set of Applications; Part 1: Functional Requirements.

European Road Transport Research Advisory Council (ERTRAC) Automated Driving Roadmap v5.0 21.07.2015.

European Road Transport Research Advisory Council (ERTRAC) Automated Driving Roadmap v7.0 29.05.2017.

German Federal Government, Strategie Automatisiertes und Vernetztes Fahren, Strategic Paper, Jun 2015.

Gräter: Rollout of automated driving in cities, infrastructure needs and benefits. Conference connected and automated driving April 2017. http://connectedautomateddriving.eu/wp-content/uploads/2017/03/Day1_B02_Graeter_EC_CAD.pdf

https://itpeernetwork.intel.com/data-center-holds-keys-autonomous-vehicles/

iMobility Forum, Automation in Road Transport, Version 1.0, May 2013.

J. Carlson, ADAS Evolving: New Developments in Hardware and Software on the Road to Autonomous Driving, HIS Automotive, Spring Media Briefing, Detroit, 19 Mar 2015.

Jean-François Bonnefon, Azim Shariff, Iyad Rahwan, The Social Dilemma of Autonomous Vehicles, In *Science*, 24 June 2016.

Nuance, DFKI: Cognitive and Conversational AI for Autonomous Driving, May 2017.

Projekt Sim-TD, Sichere intelligente Mobilität, Testfeld Deutschland, BMBF-Projekt FKZ, http://www.simtd.de

Roadmap on Smart Systems For Automated Driving, European Technology Platform on Smart Systems Integration (EPoSS) (2015).

Rudin-Brown, C. M. & Parker, H. A., 2004a. Behavioral adaptation to adaptive cruise control (ACC): Implications for preventive strategies. Transportation Research Part F, 7(2), p. 59–76.

SAE, Taxonomy and Definitions for Terms Related to On-Road Motor Vehicle Automated Driving Systems, J3016, SAE International Standard (2014).

SafeTRANS: Eingebettete Systemin der Automobilindustrie – Roadmap 2015 – 2030. 21.09.2015

SCOUT and CARTRE http://connectedautomateddriving.eu

State of California: Proposed Driverless Testing and Deployment Regulations 10.03.2017 https://www.dmv.ca.gov/portal/dmv/detail/vr/autonomous/auto

Statista: Number of cars sold worldwide from 1990 to 2017. https://www.statista.com/statistics/200002/international-car-sales-since-1990/

Tracking Automotive Technology http://analysis.tu-auto.com/

United Nations: Report of the sixty-eighth session of the Working Party on Road Traffic Safety. http://www.unece.org/fileamin/DAM/trans/doc/2014/wp1/ECE-TRANS-WP1-145e.pdf

Volkswagen: VW will enable vehicles to communicate with each other as from 2019. https://www.volkswagen-media-services.com/en/detailpage/-/detail/With-the-aim-of-increasing-safety-in-road-traffic-Volkswagen-will-enable-vehicles-to-communicate-with-each-other-as-from-2019/view/5234247/6e1e015af7bda8f2a4b42b43d2dcc9b5?p_p_auth=afnpxKt9 28.06.2017

Weltbestand an Autos: http://www.live-counter.com/autos/

Chapter 2
Applications and Use Cases

Mobility is a basic necessity for all of us. However, our current kind of mobility, using mainly vehicles, gets challenged by climate protection targets and the looming switch to renewable driving energies. And we, like anyone, cannot accept the fact that more than 1 million people get killed in traffic accidents worldwide every year. Technology alone offers options but it cannot resolve all the associated issues. Automation, electrification, networking and connectivity are key technologies to make traffic on our roads safer and more environmentally friendly. The way we move is in a state of upheaval. The first excitement around autonomous and automated vehicles was around 20 years ago and got quickly quiet. The hype around electric cars started ten years ago and subsided. But suddenly, new players out of the Silicon Valley are claiming leadership roles that were previously the privilege of Germany and German companies, mainly from the traditional vehicle and supplier industry. And in China, the market for electric cars is booming making the country the undisputed number one in production and sales.

Whether its electric mobility or self-driving vehicles - everything is going to come, maybe slowly, but surely. The computing and communications sector supplied by chipmakers, in particular, has the potential to make the vision of electrically driven, autonomous and automated driving a reality with its innovations. The scientific advisory board at the Germany's Federal Ministry of Transport has described it as the greatest disruption since the introduction of motorized road traffic. Through digitization and networking, our roles as drivers or passenger are going to change. Somehow a strange thought for a passionate vehicle driver on Germany's highways and roads.

We have watched a lot of market leaders' development efforts based upon their use cases, scenarios and application requirements for making vehicles part of the Internet of Things for many years. The vehicle ecosystem is fragmented with each stakeholder focused on specific products, applications and services. There are the vehicle OEMs and their suppliers, the data processing and service providers, the data communications and telecom operators and the semiconductor and software suppliers. Looking at them reveals that the computing and communications industry stakeholders and vehicle industry stakeholders do not always follow the same roadmaps.

We observe that vehicle drivers and passengers show new consumption patterns and mobility needs. The vehicle moves eventually from a product to buy, to a mobility service to be used, whereas the vehicles' ownership gets challenged due to the quest for efficiency. Future mobility concepts ask for drastically increased efficiency, effectiveness, flexibility and scalability. When it comes to autonomous and automated vehicles, the trend might go from focus on vehicle driving performance parameters toward the performance of infotainment, convenience, connectivity, safety, and

DOI 10.1515/9781501507243-002

security parameters. Vehicle drivers and passengers expect to have all the features and functions of their smartphone in the connected vehicle then.

Incumbent vehicle manufacturers are increasingly offering more expensive and better infotainment, telematics and assistance systems. But these excessive conservative innovations in the vehicle ecosystem open up opportunities for eager disruptors. Fully networked vehicles with open interfaces that connect ad-hoc to social networks and extract data from traffic flow and their environment are one of these looming disruptions. Automated and autonomous vehicles that warn each other, negotiate the best non-stop route, exchange data traffic perpetually with the cloud, actively react to traffic lights, signage and labelling of the road, self-steering and giving the driver and passengers the freedom to surf the internet and participate in social networks. These vehicles are going to be available for today's equivalent of 10,000 Euros or less if the customer signs for the full-service package including the mobile, entertainment, insurance and traffic flow optimization contract. The vehicle won't be the value anymore, it's the mobility and logistic service that count.

The first automated and autonomous driving series vehicles appeared in Silicon Valley, the Tesla S, as a pivoting approach to challenge the incumbent vehicle manufacturers with an electric powered vehicle. Disruptions planned around minimum viable concepts exploiting platforms like Uber, Lyft and My Taxi already change the habits in many countries. The objective is to become the world's leading platform provider of mobility services and logistics. It may create a totally new business model where the automated and autonomous driving vehicle is no longer the most important value-add, but where and on what route it moves us (and our goods) becomes the value provided by platform providers. These new disruptive business models might be the major drivers toward automated and autonomous vehicles with the required vehicle networking and connectivity thereof.

We will dive into use cases, applications and services from a timeline point of view in order to provide an insight into technology readiness and customer acceptance without raising a claim of completeness. Networked vehicles services are, for example, navigation services like traffic, speed camera location, vehicle parking space, and weather and point-of-interest travel data. Other services are in-vehicle infotainment services like web radio or video, news, gaming and access to social networks. Then there are vehicle convenience services as remote services, health reports. lifecycle management and used vehicle reports. Internet services surf via a monitor in the vehicle, read and dictate messages while driving, smartphone controls on the steering wheel while driving and streaming audio, music and video.

V2X networking and connectivity use cases have been created by many V2X ecosystem stakeholders including the computing and communications industry regulation and standardization organizations for years. It's about deriving from the use cases, V2X networking and connectivity requirements for threshold, baseline or objective values for parameters like low-latency, reliability and data throughput. Past V2X use cases can be extended by taking into account C-V2X technology-agnostic use cases with even lower latency, higher range, more advanced security and the use of broader cellular spectrum and probably increased bandwidth (Figure 2.1). The use cases spread out to non-low-latency use cases like cellular connectivity, long-range and non-active safety use cases, weather conditions, etc. as well as to non-vehicle use cases with bicycles and motorbikes, vehicle-to pedestrian (V2P), vehicle-to-home (V2H) and vehicle-to-grid (V2G), vehicle-to-network (V2N)) and infotainment.

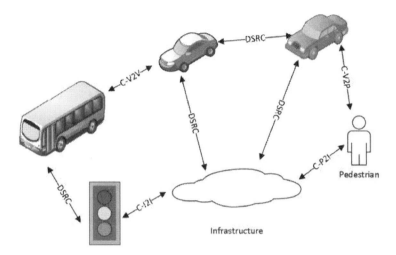

Figure 2.1: V2X system using DSRC and C-V2X

For instance, computing and communications stakeholder's use cases huddle around V2X networking and connectivity scenarios and business cases. We will also add the view of vehicle manufacturers and explore multi-stakeholder regulations and standardization organizations. To have some guidance and avoid getting lost in the overwhelming amount of use cases published from all these stakeholders, we look at V2X use cases from their time schedule and how these use cases develop over time according to the ecosystem stakeholders.

For example, vehicle-to-vehicle (V2V) and vehicle-to-infrastructure (V2I) use cases focus on the IEEE 802.11p based DSRC protocol and the dedicated ITS spectrum.

2.1 Use cases up to 2005

When it comes to V2X communications, we will start with the Federal Highway Administration (FHWA) within the U.S. Department of Transportation (DOT), which did one of the most comprehensive initial lists, summarized in Table 2.1, of V2X use cases in 2005. These use cases got categorized as use cases related to road traffic act, increasing road safety, and enabling new business models.

Table 2.1: Comprehensive use cases of the U.S. Department of Transportation (Farradyne, 2005)

Local area	Wide area
Infrastructure-based signalized intersection violation warning	Vehicles as probes for traffic data
Infrastructure-based signalized intersection turns conflict warning	Vehicles as probes for weather data
Vehicle-based signalized intersection violation warning	Vehicles as probes for road surface conditions data
Infrastructure-based curve warning	Crash data to public service answering point
Highway rail intersection	Crash data to transportation operations center
Emergency vehicle pre-emption at traffic signal	Advance warning information to vehicles
Emergency vehicle at scene warning	Electronic payment for toll collection
Transit vehicle priority at traffic signal	Electronic payment for gas payment
Stop sign violation warning	Electronic payment for drive-thru payment
Stop sign movement assistance	Electronic payment for parking lot payment
Pedestrian crossing information at designated intersections	Public sector vehicle fleet, mobile device asset management
Approaching emergency vehicle warning	Commercial vehicle electronic clearance
Post-crash warning	Commercial vehicle safety data
Low parking structure warning	Commercial vehicle advisory
Wrong way driver warning	Unique commercial vehicle fleet management
Low bridge warning	Commercial vehicle truck stop data transfer
Emergency electronic brake lights	Low bridge alternate routing
Visibility enhancer	Weigh station clearance
Cooperative vehicle-highway automation system	Cargo tracking
Pre-crash sensing	Approaching emergency vehicle warning
Free-flow tolling	Emergency vehicle signal pre-emption
Cooperative glare reduction	SOS services
Adaptive deadlight aiming	Post-crash warning
Adaptive drivetrain management	In-vehicle AMBER alert
GPS correction	Safety recall
In-vehicle signing work zone warning	Just-in-time repair notification
In-vehicle signing work highway, rail intersection warning	Visibility enhancer

Local area	Wide area
V2V cooperative forward collision warning	Cooperative vehicle-highway automation system
V2V cooperative adaptive cruise control	Cooperative adaptive cruise control
V2V blind spot warning	Road condition warning
V2V blind merge warning	Intelligent on-ramp metering
V2V highway merge assistant	Intelligent traffic flow
V2V cooperative collision warning	Adaptive headlight aiming
V2V lane change warning	Adaptive drivetrain management
V2V road condition warning	Enhanced route guidance and navigation point of interest notification
V2V road feature notification	Enhanced route guidance and navigation food discovery and payment
Rollover warning (see curve warning above)	Enhanced route guidance and navigation map downloads and updates
Instant messaging	Enhanced route guidance and navigation location-based shopping and ads
Driver's daily log	Enhanced route guidance and navigation in-route hotel reservation
Safety event recorder	Traffic information work zone warning
Icy bridge warning	Traffic information incident
Lane departure-inadvertent	Traffic information travel time
Emergency vehicle initiated traffic pattern change	Off-board navigation
Parking spot locator	Mainline screening
Speed limit assistant	On-board safety data transfer
	Vehicle safety inspection
	Transit vehicle data transfer (gate)
	Transit vehicle signal priority
	Emergency vehicle video relay
	Transit vehicle data transfer (yard)
	Transit vehicle refuelling
	Download data to support public transportation
	Access control
	Data transfer diagnostic data
	Data transfer repair-service record
	Data transfer vehicle computer program updates
	Data transfer map data updates
	Data transfer rental vehicle processing
	Data transfer video and movie downloads

Local area	Wide area
	Data transfer media downloads
	Data transfer internet audio and video
	Locomotive fuel monitoring
	Locomotive data transfer
	Border crossing management
	Stolen vehicle tracking

2.2 Between 2005 and 2011

ETSI specified a use case catalogue for intelligent transportation systems focusing on cooperative road safety, traffic efficiency, and some others that did not fit into previous ones (ETSI, June 2009). The cooperative road safety use cases are vehicle status warnings (emergency electronic brake lights, safety function out of normal condition warning), vehicle type warnings (emergency vehicle warning, slow vehicle warning, motorcycle warning, vulnerable road user warning), traffic hazard warnings (wrong way driving warning, stationary vehicle warning, traffic condition warning, signal violation warning, roadwork warning, decentralized floating vehicle data), dynamic vehicle warnings (overtaking vehicle warning, lane change assistance, pre-crash sensing warning, cooperative glare reduction) and collision risk warning (cross-traffic turn collision risk warning, merging traffic turn collision risk warning, cooperative merging assistance, hazardous location notification, intersection collision warning, cooperative forward collision warning, collision risk warning from RSU).

The traffic efficiency use cases are regulatory/contextual speed limits, traffic light optimal speed advisory, traffic information, recommended itinerary, enhanced route guidance and navigation, intersection management, cooperative flexible lane change, limited access warning, detour notification, in-vehicle signage, electronic toll collection, cooperative adaptive cruise control, and cooperative vehicle-highway automation system (platoon).

Furthermore, ETSI ITS specified use cases on point-of-interest notifications, automatic access control and parking access, local electronic commerce, vehicle rental, sharing, assignment and reporting, media downloading, map downloading and updating, ecological and economical driving, instant messaging, personal data synchronization, SOS service, stolen vehicle alert, remote diagnosis and just-in-time repair notification, vehicle relation management, vehicle data collection for product life cycle management, insurance and financial services, fleet management, vehicle software/data provisioning and updating, loading zone management, and vehicle and RSU data calibration.

In 2011, DOT focused on safety applications with connected vehicles (DOT HS 811 492A, September 2011) in a vehicle infrastructure integration architecture (Figure 2.2).

According to the DOT, these are vehicles turning right in front of buses (VTRW), forward collision warning (FCW), emergency electronic brake light (EEBL), blind spot warning (BSW), lane change warning and assist (LCA), intersection movement assist (IMA), red light violation warning (RLVW), speed compliance (SPD-COM), curve speed compliance (CSPD-COM), speed compliance work zone (SPDCOMPWZ), oversize vehicle compliance (OVC), emergency communications and evacuation information (EVACINFO), mobile visually impaired pedestrian signal system (PED-SIG), pedestrian in signalized intersection warning (PEDINXWALK), data for intelligent traffic signal system (I-SIGCVDAT), RF monitoring (RFMON), OTA firmware update (FRMWUPD), parameter uploading and downloading (PARMLD) and traffic data collection (TDC).

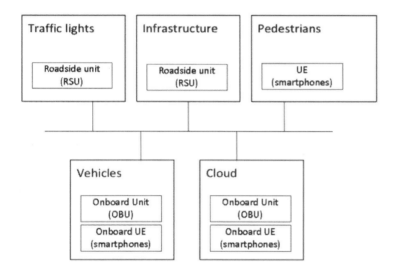

Figure 2.2: Vehicle infrastructure integration according to United States Department of Transport (DOT)

The European Automotive and Telecom Alliance (EATA) got started by the European Commission's Digital Economy and Society in September 2016 and is comprised of six associations: ACEA, CLEPA, ETNO, ECTA, GSMA, and GSA. Its main charter is to promote the wide deployment of hybrid connectivity for connected and automated driving in Europe. EATA's first objective is to see through a pre-deployment project aimed at testing the performance of hybrid communications under real traffic situations. Both EATA and 5G-PPP signed a memorandum of understanding with EATA in 2017 to collaborate on prioritization of use cases among other fields in order to better support standards for connected and automated driving. Prioritization of use cases—including future ones and the various technical requirements—needs to be agreed upon and then passed on to standards bodies such as ETSI, 3GPP, and the Society of Automotive Engineers (SAE). EATA and 5G-PPP jointly address spectrum-related issues for V2X communications and the usage modalities of certain bands, security and privacy, and vehicle safety requirements using hybrid communications (Figure 2.3).

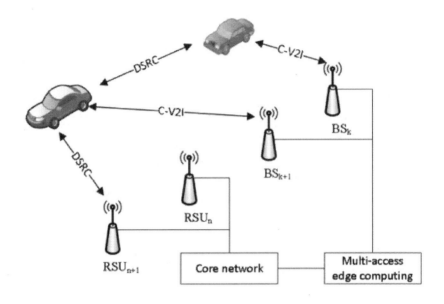

Figure 2.3: Hybrid communications for V2X according to EATA

EATA use cases are automated driving with automated overtake, cooperative collision avoidance, high density platooning, road safety, and traffic efficiency services including collective perception like see-through, vulnerable road user (VRU) discovery, bird's eye view, the digitalization of transport and logistics encompassing remote sensing and control, remote processing for vehicles and intelligent navigation, information society on the road, and nomadic nodes. During a first phase, EATA tests applications such as highway chauffeuring, truck platooning, and telecommunication network functionalities including network slicing, hybrid communications, and LTE broadcasting. The EATA aims at enhancing and upgrading the environment for existing pilot projects for the highway chauffeur Level 3 and Level 4, high-density truck platooning, and automated valet parking from 2018 onward.

The EU Commission and industry manufacturers, telecommunications operators, service providers, SMEs, and researchers initiated 5G-PPP, the 5G infrastructure public private partnership. In October 2015, 5G-PPP published a white paper (5G PPP, 2015) on its 5G automotive vision with contributions from Volkswagen, Volvo, PSA,

Bosch, Orange, Vodafone, NTT DOCOMO, Samsung, Qualcomm, Nokia, Ericsson, Huawei, CTTC, King's College London, Eurescom, and TU Dresden. The use cases and applications—including intersection collision risk warning, road hazard warning, approaching emergency vehicle warning, pre- and post-crash warning, electronic emergency brake warning, green light optimal speed advisory, energy-efficient intersection, motorcycle approaching information, in-vehicle signage, red light violation warning, and traffic jam ahead warning—were derived from European research projects simTD, DRIVE C2X, and Compass4D, completed between 2013 and 2015.

5G-PPP specifies technical end-to-end latency, reliability, data rates, communications range, node mobility, network density, positioning accuracy, and security requirements for these use cases. For example, 5G-PPP says that at least 10 milliseconds of end-to-end latency and 30 cm of accuracy are needed for automated overtake, high-density platooning, and cooperative collision avoidance. See-through and bird's eye view requires at least a 10 and 40 Mbps data rate. Vulnerable road user discovery is defined with 10 cm accuracy. These parameters are derived using end-to-end latency in milliseconds, where latency is specified as the maximum acceptable time from when a data packet is generated at the source application to when it is received by the destination application. For instance, if direct mode PC5 transport is used, this is the maximum acceptable radio interface latency. If infrastructure mode Uu transport is used, this includes the time needed for uplink, required routing in the infrastructure, and downlink.

The parameter reliability as 10^{-x} is specified as the maximum acceptable packet loss rate at the application layer after HARQ, ARQ, and so on. A packet is considered lost if it is not received by the destination application within the maximum acceptable end-to-end latency for that application. For example, a value of 10^{-5} means the application accepts at most 1 packet lost in 100,000 packets received within the maximum acceptable latency. This can be also expressed as a percentage—for example, 99.999%. Further parameters are the data rate in Mbit/s as the minimum required bit rate for the application, the range in meters as the maximum distance between source and destination of a radio link in which the application achieves a specified reliability, and the user equipment mobility in km/h as the maximum relative speed between transmitter and receiver. The network density parameter in vehicles/km^2 is defined as the maximum number of vehicles per unit area under which the specified reliability and data throughput is achieved. Positioning accuracy in centimeters is quantified as the maximum positioning error tolerated by the application. Finally, security including authentication, authenticity and integrity of data, confidentiality, and user privacy are the specific security features required by the application.

Table 2.2: Examples of use case requirements (maximum values)

Use case	End-to-end latency in milliseconds	Reliability (10^{-x})	Data rate in Mbps	Communication range in meters	Node mobility in km/h	Network density in vehicles/km^2	Positioning accuracy in centimetres
Automated overtake	10	10^{-5}	1	Up to 1000	Up to 500	Up to 3000	30
Cooperative collision avoidance	10–	10^{-5}	1	Up to 300	Up to 500	Up to 3000	30
High density platooning	10	10^{-5}	1	Up to 500	Up to 100	Up to 200	30
See-through	50	10^{-5}	10	Up to 100	Up to 100	Up to 200	30
Vulnerable road user (VRU) discovery	10	10^{-5}	1	Up to 100	Up to 50	Up to 3000	10
Bird's eye view	50	10^{-5}	50	Up to 50	Up to 100	Up to 500	30

V2X networking and connectivity characteristics are defined by the latency of 1 millisecond, broadcast/peer-to-peer/D2D, high mobility, non-line-of-sight, closing of visibility gaps, technology agnostics (802.11p, C-V2X 4G/5G), network independence or dependence, and dedicated spectrum. V2X standards applied by 5G-PPP are SAE J2735 (Dedicated Short-Range Communications [DSRC] Message Set Dictionary: intersection collision warning, emergency electronic brake lights, pre-crash sensing, cooperative forward collision warning, left turn assistant, stop sight movement assistance, lane change warning, traffic probe messages, and emergency vehicle approaching warning), IEEE P1609.1/2/3/4, ISO/IEC 8824-1/2/4, ETSI ITS, and CEN. C-V2X reuses upper layers already specified by the vehicle industry. V2X and C-V2X benefits are the support of non-line-of-sight use cases, medium range and beyond vehicle sensor reach, low-latency communications, network-independent, and all weather operation.

5GAA was formed in September 2016 by Audi, BMW, Daimler, Ericsson, Huawei, Intel, Nokia, and Qualcomm, and comprises an increasing membership including telecommunications operators. Its focus is on the development, testing, and promotion of communications solutions, their standardization and the acceleration of their global commercial availability. The 5GAA aims to address the connected mobility and road safety needs of society with applications such as autonomous driving, ubiquitous access to services, and integration into smart-city and intelligent transportation. At the Mobile World Congress in 2017, the 5GAA recorded a first list of use cases (5GAA-WG1 T-170063, April 2017) for V2X connectivity and communications together with left turn assist warning (alerts the driver as it attempts an unprotected left turn), intersection movement assist (notifies driver when it is not safe to enter an intersection), emergency electronic brake lights warning (warns the driver to brake hard in the traffic stream ahead), queue warning (provides messages and data from infrastructure of queue warnings), speed harmonization (makes available speed recommendations based on traffic conditions and weather data), real time situational awareness (delivers real-time data about city and roadway projects, lane closures, traffic, and other states), and software updates (offers mechanisms for vehicles to receive the latest software updates and security credentials).

Further use cases are remote vehicle health monitoring (delivers mechanisms to diagnose vehicle issues remotely), real-time high definition maps (provide situational awareness for autonomous vehicles), high-definition sensor sharing (provides mechanism for vehicles to share high definition sensor data from LIDAR and video cameras), see-through (provides the ability for vehicles such as trucks, minivans, and cars in platoons to share camera images of road conditions ahead) and vulnerable road user discovery (provides ability to identify potential safety conditions due to the presence of vulnerable road users). 5GAA looks to test all these use cases at V2X trials as of the RACC track (Audi, Vodafone, Huawei) at MWC 2017, ConVeX (Audi, Ericsson, Qualcomm, Swarco, University Kaiserslautern), Towards 5G (Ericsson, Orange, Qualcomm, PSA Group), Mobilifunk (Vodafone, Bosch, Huawei), UK CITE (Jaguar Land Rover, Vodafone), DT (Audi, Deutsche Telekom, Huawei, Toyota), and ICV (CMCC, Huawei, SAIC).

NGMN identifies a number of vehicle use cases in the categories of assisted driving, autonomous and cooperative driving, tele-operated driving, info-mediation, infotainment, and nomadic nodes (NGMN, September 2016). The assisted driving category includes real-time maps for navigation, speed warning, road hazards, vulnerable road users, video see-through, and sensor sharing to realize an efficient traffic flow and to reduce the number of accidents. The autonomous and cooperative driving category involves overtaking, merging, and platooning, whereas tele-operated driving covers disaster recovery, inventory, and mining. The info-mediation category contains vehicle sharing, vehicle real-time tracking, toll collecting, insurance, geo-fenced advertisement, and vehicle maintenance. Finally, the infotainment and nomadic nodes category encompass video streaming, virtual reality, augmented reality, video conferencing and in-vehicle office, as well as C-V2X relaying.

V2N connectivity and communications for dynamic high-definition digital map updates is expected to require to upload sensor data to servers with data rates up to 45 Mb/s, assuming a video H.265/HEVC HD stream with 10 Mb/s plus LIDAR data with 35 Mb/s. Data volumes to download the latest high-definition digital map information (depending on the layer details) are also estimated to be huge. NGMN assumes V2X connectivity and communications data rates for sensor sharing among vehicles and environment as of 0.5 to 50 Mbps and around 5 Mbps per link for V2X cooperation, assuming CAM/DENM and range and object detection sensor aggregation.

2.3 Use cases since 2011

The U.S. Department of Transportation's National Highway Traffic Safety Administration issued a notice of proposed rulemaking (NPRM) to mandate vehicle-to-vehicle (V2V) communications technology for new light vehicles in the United States in 2016. It addresses the use cases intersection movement assist (IMA), left turn assist (LTA), emergency electronic brake light (EEBL), forward collision warning (FCW), blind spot warning (BSW), lane change warning (LCW), and do not pass warning (DNPW). The Korean Ministry of Land, Infrastructure and Transport's cooperative intelligent transportation system (C-ITS) projects have addressed the use cases of general traffic information, speed limit information, traffic flow monitoring, local dangerous warning, speed limit warning, hard braking warning, forward collision warning, curve speed warning, blind spot warning, lane change warning, construction site warning, road condition warning, emergency electronic brake light, emergency calling, and emergency vehicle priority control. China's standardization bodies, China Communications Standards Association (CCSA), China ITS Industry Alliance (C-ITS), Research Institute of Highway (RIOH), and Telematics Industry Application Alliance (TIAA) have also been working on national, sector, provincial, and enterprise ITS standards for years.

The European CAR 2 CAR Communications Consortium (C2C-CC) was founded several years ago by European vehicle manufacturers to increase road traffic safety and efficiency by means of cooperative intelligent transport systems (C-ITS) with vehicle-to-vehicle (V2V) and vehicle-to-infrastructure (V2I) connectivity and communications. It works on solutions for the use cases of hazardous location warning, green light optimal speed advisory, approaching motorcycle warning, approaching emergency vehicle warning, warning lights on warning, roadwork warning, and traffic jam avoidance. IEEE 802.11p is proposed as the communications technology for cooperative ITS and V2V. In June 2017, the CAR 2 CAR Communications Consortium and the C-Roads Platform signed a memorandum of understanding to develop and deploy interoperable V2X-Services based on ITS-G5 on European Roads by 2019.

ISO's top-level networking vehicles use cases (ISO/TR 13185-1, 2012), (ISO/TR 13184-1, 2013), (ISO/CD TR 17427, 2013), (ISO/TR 17185-3, 2015), (ISO 17515-1, 2015), (ISO/TS 16460, 2016), and (ISO 13111-1, 2017) are maintenance, insurance, infotainment, in-vehicle commerce, telematics, C-V2X automated driving, and smart cities. Maintenance comprises remote monitoring, predictive maintenance, and FOTA/SOTA. Insurance comprises PAYD and PHYD. Infotainment comprises mobile broadband, augmented navigation, and in-cabin streaming. In-vehicle commerce use cases comprise, for example, parking, toll and software upgrades. Telematics comprise crash management, theft monitoring, remote entry, and emergency call. V2X automated driving comprises awareness driving, sensing driving, cooperative driving, and autonomous driving. Smart cities comprise traffic management, electronic tolling, road conditions, smart-home controls, and eco-driving rebates.

3GPP TS 22.185 (3GPP TS 22.185, 2017) specifies V2X, comprised of the types: vehicle-to-vehicle (V2V), vehicle-to-infrastructure (V2I), vehicle-to-network (V2N), and vehicle-to-pedestrian (V2P). Basic application classes are road safety, traffic efficiency, and others. 3GPP TR 22.885 specifies use cases, requirements including safety and non-safety features for LTE to support V2X services taking into account inputs from other SDOs—for example, GSMA, ETSI ITS, U.S. SAE, or C-ITS. Major identified uses cases are V2V connecting vehicles, V2P linking vehicles and pedestrians or others, and V2I networking vehicles with roadside units, all based on LTE communications. In addition to the use cases 5G-Americas references, 3GPP TR 22.885 Release-14 V2X (3GPP TR 22.885, 2015) contains the use case of message transfer under mobile network operator (MNO) control, V2X in areas outside network coverage, V2X road safety service via infrastructure, pedestrian-against-pedestrian collision warning, V2X by UE-type RSU, V2X minimum QoS, V2X access when roaming, pedestrian road safety via V2P awareness messages, mixed use traffic management, privacy in the V2V communications environment, providing overviews to road traffic participants and interested parties, remote diagnosis, and just-in-time repair notification.

3GPP TR 22.886 Release-15 (3GPP TR 22.886, 2015) use cases comprise eV2X support for vehicle platooning (Figure 2.4), information exchange within platoon, vehicle sensor and state map sharing, eV2X support for remote driving, automated cooperative driving (Figure 2.5) for short-distance grouping, collective perception of environment, communications between vehicles of different 3GPP RATs, multi-PLMN environment, cooperative collision avoidance (CoCA) of connected automated vehicles, information sharing for partial and conditional automated driving, information sharing for high and full automated driving, information sharing for partial/conditional automated platooning, information sharing for high/full automated platooning, dynamic ride sharing, multi-RAT, video data sharing for assisted and improved automated driving (VaD), changing driving-modes, tethering via vehicle, out of 5G coverage, emergency trajectory alignment, tele-operated support (TeSo), intersection safety information provisioning for urban driving, cooperative lane change (CLC) of automated vehicles, secure software update for electronic control unit, and 3D video composition for V2X scenario.

Figure 2.4: Vehicle platooning

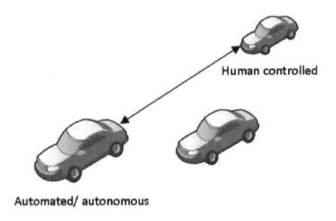

Figure 2.5: Automated or autonomous cooperative driving

3GPP TR 22.891 (3GPP TR 22.891, 2015) studied use cases for Release 14 to identify re-lated high-level potential requirements for 5G in 2016. V2X related use cases are ultra-reliable communications, network slicing, lifeline communications and natural dis-aster, migration of services from earlier generations, mobile broadband for hotspots scenarios, on-demand networking, flexible application traffic routing, flexibility and scalability, mobile broadband services with seamless wide-area coverage, virtual presence, connectivity for drones, tactile internet, localized real-time control, coex-istence with legacy systems, extreme real-time communications and the tactile internet, remote control, lightweight device configuration and wide area sensor monitoring, and event driven alarms.

5G Americas was founded in January 2002 to unite wireless operators and vendors in the Americas working with regulatory bodies, technical standards bodies, and other global wireless organizations. 5G Americas references 3GPP for V2X use cases (5G Americas, April 2017)—in particular, 3GPP TR 22.185 and 3GPP TR 22.886. The 3GPP Release-14 use cases are forward collision warning, control loss warning, emergency vehicle warning, emergency stop, cooperative adaptive cruise control, queue warning, road safety services, automated parking system, wrong way driving warning, pre-crash sensing warning, traffic flow optimization, curve speed warning, vulnerable road user safety, and enhanced positioning. The 3GPP Release-15 and beyond use cases are vehicle platooning, sensor and state map sharing, remote driving of vehicles (Figure 2.6), and collective perception of the environment.

V2X server

Figure 2.6: Remote driving

In 2017, there were further announcements from vehicle manufacturers about their latest connected vehicles or their succeeding generations of autonomous driving vehicles. Together with mobile network operators, these vehicles get tested in trials. For instance, Deutsche Telekom in Germany runs trials of V2X connectivity and communications on the Ingolstadt autobahn test bed in Germany, together with Huawei, Audi, and Toyota. Verizon in the United States has dealt with V2X use cases for some years. The operator was part of the M City project, which was an eight-year project from the United States Department of Transportation and a consortium of corporate partners including Ford, GM, Honda, Nissan, and Toyota. Suppliers were Delphi, Denso, Bosch, Qualcomm, and others. Verizon is engaged in V2V trials in Ann Arbor, Michigan at the Ann Arbor Connected Vehicle Test Environment (AACVTE).

In Japan, NTT DOCOMO works together with Continental to enhance connected infotainment functions and build the first solutions for cellular-based V2X wireless communications systems. In China, China Mobile, Huawei, SAIC Motor, and Xihu Electronics demonstrated C-V2X use cases including bus and vehicle interactivity, see-through, driving guide for traffic lights, alarms for pedestrians, change lanes, and emergency brakes and alarms in 2016. China Mobile has C-V2X test beds with SAIC, Huawei in Shanghai that have remote drive use cases using multiple video cameras in the vehicle with a 240-degree view of the vehicle's surroundings and control signals for the steering wheel, gas pedal, and brakes. Another test bed, intelligent vehicle integrated systems test area, with Changan and Datang is in Chongqing providing 11 types of road and 50 driving scenarios. Further test beds are with Baidu, ZTE, Dong Feng, and Huawei in Beijing and with Wuhan in Changchun.

The vehicle manufacturers are certainly the frontrunners when it comes to scenarios and use cases for V2X. We are already exposed in today's traffic to many very different scenarios like dense traffic, monotonous long travel and poor visibility just to name a few challenging ones. Vehicle manufacturers' advanced driver-assistance systems (ADAS) support us already being the driver in many of these stressful scenarios and use cases today. First, we give an overview and discover the safety and infotainment related uses cases supporting you in your vehicle today and their evolution in the near term. Second, we look at the technology providers for networking and connectivity to compare their technical point of view with the ones of the vehicle manufacturers.

Many vehicles have offered comfort and safety-enhancing driving assistance systems as a first key aspect as a vehicle standard for quite some years. These systems are classified as driving assistance systems for example for urban, rural and parking scenarios. The urban scenario encompasses use cases similar to side assist, cross traffic warning, vehicle exit assist and trailer assist. The rural scenario contains adaptive cruise control (ACC), navigation, traffic jam assist, pre-sense front, active lane assist, turn assist, light assist, traffic sign recognition and predictive efficiency assist use cases. For example, the pre-emptive pre-sense city system in the Audi Q7, where a front camera detects an imminent collision, warns the driver and initiates a full braking if necessary.

Another urban scenario use case implementation example is the Audi V2I traffic light information (TLI). It signals real-time data from the advanced traffic management system that monitors traffic lights via an on-board 4G LTE data connection and is now available for Audi A4, Q7, and A4 all-road models. The traffic signal data get displayed on an instrument cluster or HUD as time-to-green countdown. Future services will be the integration into vehicle start and stop features, navigation guidance and route optimization, and predictive services like green light optimized speed advice (GLOSA). The use case is implemented in partnership with the Traffic Technology Services (TTS), Regional Transportation Commission of Southern Nevada (RTC) and city of Las Vegas. General Motors demonstrated as well with the Cadillac CTS vehicle-to-infrastructure (V2I) capability whereas vehicles receive real-time data from traffic light controllers on signal phasing and timing in 2017.

The current road traffic act and safety related use cases do not make use of V2X networking and connectivity so far. But these use cases will evolve. Millions of vehicles have been equipped with emergency call services for years providing a wireless link into the vehicle. Typical additional use cases of these V2I application are emergency electronical brake light (EEBL), forward collision warning (FCW), intersection collision warning (ICW), stationary vehicle warning (SVW) and pre-crash-warning (PCW). We find in the vehicle automatic high beams, forward collision warning, front automated emergency braking, lane departure warning, lane keeping assist, blind-spot monitor, front parking assist, parking assist, rear automated emergency braking, rear cross-traffic monitor, rear parking assist, automatic distance control (ADC), blind spot sensor, lane assist, light assist, dynamic light assist, drowsiness warning, side assist, traffic sign recognition and traffic jam warning (TJW).

And now V2X networking and connectivity moves in. For example, an ITS safety package is available on three Toyota models in Japan for the 2017 Prius, Lexus RX and Toyota Crown. It is based on standardized ITS/DSRC frequency of 760 MHz and enables V2V applications like radar cruise control, emergency vehicle notification and V2I applications as of right-turn collision caution, red light caution and traffic signal change advisory. In the United States and Canada, the General Motors' new Cadillac CTS comes equipped with Cohda's V2X communications based on the IEEE 802.11p in 2017. The data exchange with a range of up to 300 m and up to 1,000 messages per second shall increase safety and efficiency in road traffic. The new Cadillac therefore accesses data about traffic on other vehicles and infrastructure that are present in the current environment. The system supports the use cases collision warning, location and emergency warning. Several other use cases will be supported by General Motors' hands-free, highway-driving system called Super Cruise, which is planned for the Cadillac CT6 luxury sedan in 2017.

ADAS related there are the most advanced use cases of autonomous driving (measure and synchronize driving trajectories by in-vehicle sensors) and automated driving. Here V2X communications augments driving beginning with basic warning messages that require driver intervention up to the increased levels of automation which are enabled for instance by sharing of sensor data and trajectories. Sensing driving is an inherent necessary part of these where the vehicles broadcast data gained through on-board sensors (camera, radar, LIDAR, and ultrasound) to neighboring vehicles as well as to the infrastructure and the cloud. The cloud feeds back real-time mapping and traffic update data to specific vehicles. Now vehicles see with the eyes of other vehicles either directly (V2V) or via the cloud (V2I). So, vehicles detect otherwise hidden objects e.g. around the next building and get a more extended view (eHorizon) on what is happening within the vehicle surrounding. Use case examples include overtaking warning and intersection collision warning.

The electronic horizon provides other ECUs a continuous forecast of the upcoming road network by using optimized transmission protocols. It integrates map matched localization and positioning, most probable path, static map data like curvature, slopes, speed limits and road classification as well as dynamic map data like route, traffic, hazard warnings, road construction status and weather. These data are input to the ADAS applications resembling automated and autonomous driving, fuel efficient driving, predictive curve light or curve speed warning. Therefore, map and navigation data become additional sensors for ADAS. The standard electronic horizon transmission protocol for the communication between maps and ADAS is developed by the ADASIS consortium. The navigation maps as electronic horizon supplier, require updates and more and more connectivity with rising SAE levels.

The solutions for driving safety and ADAS like parking assist, traffic jam assist, road sign assist, remote park assist, side, intersection movement assist, rear-view assist, lane change assist, electronic emergency brake light, predictive emergency braking, predictive pedestrian protection, intelligent headlight control, lane departure warning, lane keeping support, vehicle dynamics management, vehicle electronic stability, and driving comfort aggregate data from a number of sources, including camera, radar, and ultrasonic sensors and are delivered by suppliers like Bosch, Continental, Delphi and Harmann (now Samsung). The solutions add functions for connectivity control and eCall, cellular wireless, Wi-Fi, navigation, infotainment, software services and apps, HUD, dual view display, and programmable instrument clusters. The solutions for vehicle connectivity and communications may already include body (sensing, charging, monitoring) and security functions (anti-theft sys, interior intrusion detection).

An intermediate stage is the use case of awareness driving, where vehicles disseminate their status data like position, speed, and direction to all notice taking vehicles. It enables them to become aware of vehicle presence and of eventual hazards on a need to know basis. These data include pre-crash warning, vehicle control loss warning, emergency vehicle warning, emergency stop warning, curve speed warning, intersection safety warning and queue warning. The telematics control unit (TCU) or user equipment (UE) relay warnings to the driver or to in-vehicle ADAS/AD systems. Another "in-between" use case is cooperative driving. Here vehicles share certain data with other traffic participants. These data are fed to the vehicle's automated driving algorithms to accurately anticipate what other traffic participants will do next. Examples include platooning, cooperative collision avoidance and automated overtake.

A glimpse into the evolution of V2X use cases give for instance the trials supported by all V2X ecosystem stakeholders. With the 2018 level-3 autonomous driving Audi A8, a solution for piloted driving up to 60 kilometers per hour for use cases like traffic jam pilot and remote park pilot is provided that uses today's available technology. It is expected that by the beginning of 2020, vehicles will drive themselves up to 100 kilometers per hour under highway conditions, including lane changes, passing, and responding to unexpected scenarios. BMW explores 5G V2X use cases in partnership with SK Telecom and Ericsson in 2017. Applications are video recognition and obstacle detection with the exchange of safety data between vehicles, real-time multiview streaming, 4k 360 VR surround view, HD live conference system and drone helper with a high-quality bird's-eye view for the driver. The technology used is a millimeter-wave 28 GHz radio access network on a 2.6 kilometer test track at BMW driving center in Yeongjongdo, Incheon, Korea providing 3.6 Gbps at 100 kilometer per hour.

Daimler demonstrates the driverless parking use case in real-life traffic in the multi-level vehicle park of the Mercedes Benz Museum in Stuttgart in 2017 and has been shown S- Mercedes luxury vehicles and trucks autonomously driven autonomously since years. The Mercedes-Benz S- and E-class vehicles reach with their Intelligent Drive driving assistance solutions advanced stages of automation. Among the available supported use cases, which also lead Daimler step by step closer to the goal of driverless driving are: active speed limit, active distance assist, active steering assist, evasive steering assist, route-based speed adjustment, active emergency stop assistant, semi-automated driving on freeways, highways and in city traffic, autonomous braking, active brake assist cross-traffic, active lane keeping assist and active blind spot assist. And PSA shows a Citroën C4 Picasso from its autonomous vehicles fleet driving autonomously through a toll station in Paris using communications between the vehicle and the infrastructure in 2017.

Suppliers maintain several ecosystem partnerships for V2X trials too. For instance, Continental and NTT DoCoMo have got a collaboration on 5G-communications for road traffic systems with the objectives to integrate infotainment functions and V2X communications by 5G in 2017. V2V without network involvement is going to be used for applications queue warning and left turn assist. 5G is then expected to support low latency and direct communications use cases in almost real-time like see-through sensor sharing (Figure 2.7). Wireless communications technology suppliers partner with several vehicle manufacturers conducting field trials. The objective is to test use cases for vehicle connectivity and communications like see through between two connected vehicles and the emergency vehicle, which aims at notifying drivers when an emergency vehicle is approaching. Both cases exploit the network-based capabilities of V2X connectivity and communications to deliver a high-resolution video stream between two vehicles.

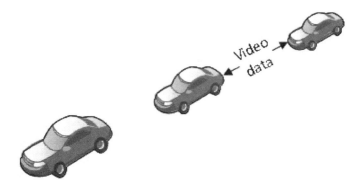

Figure 2.7: Sensor sharing between vehicles

V2X is seen as a critical component for autonomous and automated driving to support non line-of-sight sensing by providing 360° NLOS awareness for instance at intersections, on-ramps, at rain, fog and snow environmental conditions, at blind intersections and for vulnerable road user (VRU) alerts. V2X connectivity and communications enables conveying intent and share sensor data to provide high level of predictability for example for road hazard and sudden lane change warning. V2X contributes to situational awareness offering increased electronic horizon to enable soft safety alerts and reliable graduated warning for reduced speed ahead warning, queue warning and shockwave damping. In particular V2V supports collision avoidance safety systems, V2P safety alerts to pedestrians and bicyclists, V2I traffic light optimal speed advisory and V2N real-time traffic routing and cloud services. Enhanced safety use cases extending the vehicles' electronic horizon, providing more reliability, and better NLOS performance. They are for example: do not pass warning (DNPW), intersection movement assist (IMA) at a blind intersection, blind curve and local hazard warning, vulnerable road user (VRU) alerts at blind intersection, road works warning and left turn assist (LTA). Finally there are trials that use IEEE 802.11p like Korea's Cooperative Intelligent Transportation System (C-ITS) project initiated by the Korean Ministry of Land, Infrastructure and Transport with an 87.8 km route between Sejong City and the Shanghai Intelligent and Connected Vehicle Demonstration Program.

Infotainment, human-machine-interface (HMI) and telematics are the second focal point which we see when it comes to V2X scenario and use cases. In-vehicle infotainment related are use cases like augmented navigation, mobile broadband and in-cabin streaming. A wide range of HMI solutions where the driver can use a touchscreen, a central operating panel, the steering wheel remote control, or natural voice input to control the different functions for audio, video, navigation, communications, or other vehicle's convenience technologies including in-vehicle connectivity like CAN, LIN, FlexRay, and Ethernet to link embedded control units is provided by suppliers like Bosch.

Augmented navigation is a major complementary part of infotainment and enhances traditional 2D or 3D navigation apps with real time traffic camera feeds from intersections, roadside infrastructure or other vehicles. An outcome for example is a multi-layered high-resolution local dynamic map (HD-LDM). The HD-LDM layer one covers static data from the map vendor. Layer two encompasses traffic attributes, static road side units and communications node data, intersection data and landmarks for referencing and positioning. Layer three consists of temporary regional data like weather, road or traffic conditions. And layer four incorporates dynamic communications nodes data as well as other traffic participants detected by in-vehicle sensors. Mobile broadband includes applications that require high-bandwidth wireless broadband connectivity and communications such as rear seat 4K or 8K video streaming, virtual reality (VR) cloud gaming. In-cabin streaming is a bi-directional streaming of media content from the vehicle infotainment platform and brought-in devices like smartphones over the in-vehicle wireless network using for instance Wi-Gig. There are by now plenty of infotainment and ADAS applications in relationship with in our vehicles. These are between navigation and maps, adaptive cruise control, speed limit assist, take-over assist, lane change assist, active lane assist, cross traffic warning, rear and front collision warning, wrong driver warning, park assist, remote parking, active park distance control, rear and surround view and night vision with people recognition. For example TCU modules from Harmann are ready for V2V and V2I including eCall, real-time traffic reports, service bookings up to cloud-based analytics and server platforms and evolve recently into ADAS. These modules include a complete multi-layer security architecture solution using five security features in conjunction with OTA updates to optimize protection.

Business related use cases as third main emphasis seen in the area of V2X networking and connectivity. Services for intelligent transportation systems (ITS) where ITS back end systems pull data from V2X capable TCUs for traffic management including remotely controlled dynamic routing, dynamic electronic toll collect and insurance or predictive vehicle maintenance are seen as emerging business opportunities. Traffic management including road conditions exploits sensor data from vehicles to optimize the traffic flow and to report and take action on road conditions in case of a smart city. Examples consist of variable message signs with real time traffic or routing info, dynamically updated traffic lights and updating city authorities of pot holes or on poor road conditions. Dynamic electronic toll collect stretches from pre-paid or real-time toll collect over re-configurable and usage based vehicle insurances up to a seamless pay-as-you-go parking, where a vehicles reserves a parking lot in a multi-floor park house and initiates a payment pre-authorization. Usage based vehicle insurance uses customized driver analytics data stored to the cloud and compared in real time with its behavior in current traffic to adjust insurance fees. And an eco-driving rebate offered by green smart cities awards for driving eco-friendly.

For example the remote and predictive vehicle maintenance use case analyses in-vehicle ECU and other data transmitted to the cloud for deviations from the norm. The data gathered to improve the knowledge base to predict upcoming failures on vehicles that show the same deviations and support edge and cloud predictive analytics. In edge predictive analytics, the database of vehicle reference models is deployed at the network edge or vehicle itself. An algorithm tests degradation of monitored parts against the models and notifies of impending part failure. In cloud predictive analytics the reference models database is maintained in the cloud. This maintenance use case assumes a secure update of firmware and software, where the vehicle is in a safe state, and the ECUs or TCUs process the update and restore the vehicle to a working state at completion of update.

Computing and communications technology suppliers have been part of the V2X ecosystem for years and provide their view on scenarios and use cases. In 2017 we recognize three major platforms for V2X networking and connectivity solutions paving the way towards automated and autonomous driving capabilities. First, there are platforms which gather, process and analyse data from autonomous or automated vehicles and transmit and store them in data servers. This end-to-end solution has the capabilities to enable vehicle manufacturers to do their own data analytics and deep learning in their own data centers. A second connected vehicle platform option follows a connectivity and communications centric path towards autonomous and automated driving and provides dedicated solutions with plenty of experience in telematics and V2X connectivity and communications. There is no one data center solution yet, but the solution exploits the advantage of being available today in in-vehicle infotainment (IVI) and connectivity and communication solutions like vehicle grade LTE with the capability to communicate between vehicles and the network.

The third platform option works on a slightly different unique approach for autonomous and automated driving which exploits parallel computing strength and is more focused on artificial intelligence (AI) and machine learning. The solution uses graphic processing units (GPUs) in the data center to train neural networks offline and in the vehicle to do inference. This platform assumes not a full V2X connectivity for autonomous or automated driving. But V2X communications could be added for instance to do updates over-the-air (OTA). The supported use cases are centered around this platform focus on sensor processing and fusion, detection and classification, localization in particular HD map localization and interfacing, vehicle control, scene understanding and path planning and finally streaming to computing clusters.

The evolution of all these platform options towards automated and autonomous driving is strongly linked with 5G communications technology development supporting SAE automation level scenarios and use cases. It starts with supporting the no automation level (status data like I'm a vehicle at coordinates x, y and z, traveling West at 80 kilometers per hour), driver assistance level (sensor data like status data and it's raining at my location and I just passed a VRU), partial automation level (intention data like status plus sensor data plus merging, coming along side), conditional automation level (coordination data plus sensor data plus intention plus let's platoon and slot based intersections) and to end with high and full automation level with massive data sharing and sensor data fusion.

The current V2X standards DSRC and ITS-G5 offer a foundation for basic safety scenarios and use cases such as forward collision warning implementing 802.11p as physical radio layer. 3GPP C-V2X Rel-14 supports enhanced safety use cases at higher vehicle speeds and challenging road conditions requiring improved reliability, extended range, low latency and non-line-of-sight (NLOS). C-V2X supports direct communications operating in the ITS 5.9 GHz band without network assistance, which makes it an option for vehicle-to-vehicle (V2V), vehicle-to-infrastructure (V2I), and vehicle-to-pedestrian (V2P) communications. Vehicles communicate with each other and roadside infrastructure without requiring a subscriber identity module (SIM), cellular subscription or network assistance. Disabled vehicle after blind curve, and do-not-pass warning and road hazard warnings in varying road conditions are examples of these enhanced safety use cases.

C-V2X has an evolution path toward 5G including backward compatibly addressing additional use cases and safety requirements. The first phase establishes a foundation for basic V2X use cases like forward collision warning and basic safety with 802.11p or C-V2X 3GPP Rel-14. The second phase exploits a better communications link budget providing longer range and increased reliability with C-V2X 3GPP Rel-14 for enhanced safety use cases like disable vehicle after blind curve. And the third phase is going to enable advanced safety use cases like see-through and video, radar or LIDAR sensor sharing, cooperative driving or bird's eye view of intersection approaching vehicles and 3D HD map updates with C-V2X 3GPP Rel-15 or 16.

The evolution of C-V2X toward a support of automated and autonomous driving use cases holds the promise to fulfill the requirements with 5G networking and connectivity. Current autonomous and automated vehicle prototypes work independently from cellular networks in the common sense in that they find their way with available on-board communications and computing power. But without V2X connectivity and communications these vehicles in large numbers make the same problems that individual drivers produce today. V2X connectivity and communications is needed requiring a highly flexible, adaptive, and reliable and low-latency network. With short and long range, high bit rates, very low latency and enhanced security, today 5G promises exactly what is needed. And 5G shall complement existing communications technologies in the vehicle sector, such as the ETSI ITS G5 intelligent transport system proposal.

Wireless infrastructure suppliers expect that 5G-enabled automated driving will not happen suddenly and there is a lot that can be used from 4G LTE and its evolution towards 5G. Since V2X is about vehicles, it includes other vehicles like excavators and mining vehicles as well and even a bus can be computerized or automated further. Therefore, wireless infrastructure providers elaborate on use cases like assisted driving, which warns road users in advance about road hazards or helps when overtaking other vehicles as well to provide e.g. latency-optimized protocol stacks and dynamic edge cloud computing supporting fast mobility.

The use cases evolve over time on the way to 5G networks and may require even more networking and connectivity functions. It should be emphasized that ETSI ITS G5 technology alone might not be enough to meet these requirements and LTE complements ITS G5 for V2V and V2I communications as direct V2V communications for proximity, path prediction and collision anticipation and warning, advanced driving assistance with collective perception, sensor sharing, cooperative lane change, traffic safety with vulnerable road user protection, intersection assistance, accurate positioning, platooning in particular see-through, intersection and lane change as well as rear end warning use cases. LTE and the future 5G does mid and long range V2X connectivity and communications in case of the weather, road and traffic conditions electronic horizon use case. And the wireless infrastructure supplier's 5G V2X use cases embrace cloud based infotainment with broadband multimedia, update lane-level maps with real time context and V2V communications for increased road safety and comfort (e.g. truck platooning) and vehicle analytics with real time data analysis as well.

Whereas wireless infrastructure suppliers presume for in-vehicle infotainment, high definition maps, location based services and vehicle maintenance no ultra-reliability nor low latency though it does anticipate an increasing demand on vehicle safety due to autonomous and automated driving with an increasing need for low latency and reliability of connectivity and communications. In particular with autonomous driving, the V2V and V2I communications has to be fast and reliable as it will be used for example to broadcast warning messages. Accurate positioning and high availability of the communications both in time (end-to-end (E2E) latency below 5 milliseconds as application level delay) and space (positioning accuracy shall be less than 0.5 meters) are needed as well. The availability is aiming for 100%. The vehicle communication service should be ubiquitously available. This is not equivalent to 100% mobile coverage as the V2X communications is ubiquitous by itself. Wireless infrastructure suppliers see V2X connectivity and communications value first in infotainment and mapping and navigation. Second, network function virtualization (NFV) and V2X plus cloud open opportunities for evolving telematics services like insurance, maintenance and traffic management. And third, V2X has to evolve to be useful for automated and autonomous driving.

2.4 Conclusions

The role of V2X connectivity and communications on our way forward toward autonomous and automated driving is mainly determined to reduce traffic fatalities with highly optimized passive safety and transport efficiency by highly automated vehicles (HAV) with increased situational awareness, provision for proactive driver warnings and intervention to prevent and mitigate crashes where driver response is late or non- existent. The increasing number of sensors in highly automated vehicles and V2X connectivity and communications is going to enable steady rising levels of automation corresponding to the BASt, NHTSA and SAE levels of automation. Today's V2X networking and connectivity use cases implementing the road traffic act, increasing road safety and enabling business cases as shown in Figure 2.8 have been around for quite a lot of years. Therefore, we classify three main use case categories, which are road safety (ADAS), in-vehicle infotainment and services (business model) related. Nevertheless, we still witness some issues and opportunity for improvements.

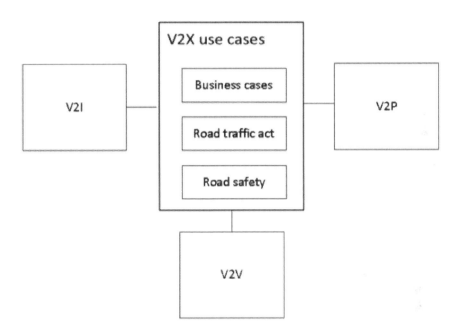

Figure 2.8: Use case categories supported by V2X networking and connectivity

First, the use cases from all ecosystem stakeholders do not vary significantly according to the categories done years ago, even if they are called traffic efficiency, co-operative road safety and infotainment now, like the Table 2.3 below shows exemplarary.

Table 2.3: Use cases according to road traffic act, road safety, and business cases

Road traffic act	Road safety	Business cases
Active bending headlights	Airbag control	Active safety and insurance
Adaptive cruise control	Anti-lock braking system or anti-skid braking system	Assistance call
Adaptive lighting	Automatic emergency braking	CO_2-aware driving
Approaching emergency vehicle warning	Automatic vehicle stop	Cooperative vehicle-highway automation system
Auto park/park assist	Backup assist	Electronic toll collection
Automated overtake	Bad weather warning	Energy-efficient intersection
Bird's eye view	Blind merge warning	Fleet management public or private
Bus priority signal control	Blind spot detection	Freight as a service
Cooperative adaptive cruise control	Blind spot warning	Infotainment
Cooperative collision avoidance	Collision imminent steering	Insurance
Cross-traffic warning	Collision risk warning	Insurance black box
Do-not-pass warning	Control loss vehicle warning	In-vehicle signage
Electronic emergency brake warning	Cooperative forward collision warning	In-vehicle video camera to record accident incidents
Green-light optimum speed advisory	Cyclist detection	Just-in-time repair notification
High density platooning	Driver behavior profile broadcast	Multimodal transport and logistics
Highway assist	Dynamic vehicle warnings	Navigation maps
Highway chauffeuring	Emergency call (eCall)	Navigation service
Highway merge assistant	Emergency electronic brake light	Remote diagnosis
Intelligent high-beam control	Emergency vehicle signal pre-emption	Remote maintenance
Intersection collision risk warning	Fatigue warning device	Safety recall notice
Intersection movement assist	Forward collision warning	Smart cities
Lane change assist	Hard braking ahead warning	SOS services
Lane departure warning	Intersection collision warning	Stolen vehicle tracker
Lane keep assist	In-vehicle AMBER Alert	Telematics
Left turn assist	Large animal detection	Traffic management
Low bridge/bridge height warning	Motorbike detection	Vehicle and mobility as a service
Low parking structure warning	Pedestrian detection	Vehicle sharing (public or private)

Road traffic act	Road safety	Business cases
Night vision	Post-crash warning	
OTA firmware update	Pre-crash sensing	
Oversized vehicle warning	Real-time situational awareness	
Pedestrian crossing information at intersection	Rear cross-traffic alert	
Queue warning	Reverse camera	
Rail collision warning	Side blind zone alert	
Railroad crossing violation warning	Slow/Stopped vehicle warning	
Red light violation warning	Stop sign violation warning	
Reduced speed zone warning	Surround view camera	
Remote sensing and control	Traffic hazard warnings	
Right turn assistant	Traffic jam ahead warning	
Road condition warning	Trailer sway control	
Road hazard signalling	Vehicle status warnings	
Road hazard warning	Vehicle type warnings	
See-through	Vehicle-based road condition warning	
Speed compliance work zone	Vehicle-to-vehicle road feature notification	
Speed harmonization	Visibility enhancer	
Spot weather information warning	Vulnerable road user detection	
Stop sign gap assist	Wrong way driver warning	
Stop sign movement assistance		
Traffic jam ahead warning		
Traffic jam assist		
Traffic sign recognition		
Traffic signal violation warning		
Transit and freight signal priority		
Truck platooning		
Weighing-in-motion		
Work zone/road worker warning		

Second, the majority of use cases stem from the communications and computing stakeholders, whereas the vehicle manufacturers hold back. Nevertheless, vehicle manufacturers have got plenty of use cases already addressed by their vehicles' ADAS systems. Additionally, V2X connectivity and communications are already implemented for infotainment and navigation use cases and are ready to go with the availability of communications modems and smartphones built into the vehicle for many other business cases (including insurance). What role does V2X play for road traffic acts and road safety use cases?

One role is certainly the possible extension of the active safety limit beyond the line-of-sight—for example, with cooperative adaptive cruise control and intersection collision warning. Under the assumption that V2X provides safe and secure connectivity and communications, V2X supports a distinctive number of road traffic acts and road safety use cases that cannot otherwise be covered by vehicle sensors. Some examples are non-line-of-sight intersection movement assists, blind spot detection, lane departure and lane change warnings, left-turn assists, or emergency electronic brake lights. Another scenario in which V2X might apply advantageously is when the vehicle on board sensor range is too short, or when sensor data are sacrificed due to adverse weather and environmental conditions. The fusion of V2X connectivity and communications with in-vehicle sensors gives—under certain previously stated assumptions—increased reliability, increased accuracy, improved warning timing and redundancy, and backup in case of failures. V2X could alert and warn complimentary drivers and vehicles themselves similar to ADAS alerts in use cases such as automatic braking, steering, and parking, or cooperative adaptive cruise control and, ultimately, autonomous and automated driving.

Another role of V2X is being an enabler of wider intelligent transportation systems (ITS) solutions within the stakeholder ecosystem and its evolution into a pervasive, cross-vertical connectivity and communications service for the Internet of Things. The debate around inevitability of V2X is still ongoing and open, but generally, V2X connectivity and communications is considered critical for Level 4 autonomous—and certainly Level 5 automated—vehicle operation. To achieve this goal, V2X connectivity and communication has to be incorporated into a wider, unified autonomous and automated vehicle framework to support data gathering, fusion, storage, and collaborative processing. Assuming a successful integration, this framework then enables much wider cooperative, shared, and service-based business cases for current vehicle sharing such as vehicle-as-a-service (VaaS), mobility-as-a-service (MaaS), freight-as-a-service (FaaS) and smart cities or IoT paradigms such as vehicle-to-grid (V2G) and vehicle-to-home (V2H).

Third, we do see room for improvement for the current use cases by adding more the real world's safety and security requirements. What do we mean by that? The road to the future of autonomous and automated vehicle driving goes straight through the big megacities in Asia, America, Australia, and Europe, with their unpredictable and chaotic urban environments. If V2X connectivity and communications wants to be successful, the megacities are proving grounds for level 3, 4, and 5 vehicles with V2X. The real world of a megacity looks a little bit different compared to the cleaned up use cases or test trials for autonomous and automated driving.

At one moment, a siren of an emergency vehicle will swell, a delivery truck will block our escape way, and the drivers behind will immediately honk at us. After we finally escape at the next corner, a pedestrian exits a taxi cab on the street without looking around, and someone on a bike quickly comes at us from the wrong way before we hit the brake for a full stop. Does this sound familiar? Driving in a megacity requires every little bit of attention we can gather. Such a scenario happens every second in the global megacities, and is very different in complexity compared to the level 3 or higher vehicle test-driving on highways or less crowded city centers. Of course, there is the use case of the autonomous driving mode and safely navigating all kinds of weather on a well-marked highway. But we need use cases that cover the extreme nature of a rural farm, as well as one for a dense urban megacity under all environmental conditions. V2X has to work reliably, safely, and securely under all conditions, complexities, and extremes to be of value for SAE's Level 3, 4, or 5 automation level. And that is by no means an easy task.

The evolution of V2X connectivity and communications is strongly interconnected with the different levels of vehicle automation. It is clear that level 5 fully automated vehicles require significantly greater data sensing, storage, processing, and communication capabilities than level 3 or level 4 vehicles. Data have to be processed and communicated reliably, securely, and safely in real-time. These needs are not very well reflected by V2X networking and connectivity technology today in the use cases looked at, raising concerns that today's existing V2X technology might need some more development to meet all of these needs. But we believe that no one has been able to demonstrate a V2X technology that meets the needs of the vehicle targeting real-world megacities yet. It will take the entire power and resources of all V2X ecosystem stakeholders to evolve from Level 3 vehicles to Level 4 and 5 to finally set up the required V2X connectivity and communications framework that will finally fulfil the autonomous and automated vehicle driving promises.

V2X networking and connectivity might evolve into an IoT-like, ubiquitous, horizontal connectivity solution which is used by many vertical applications and services. There is a common agreement among all stakeholders that V2X networking and connectivity has to be considered as critical for SAE level 4 autonomous and definitely level 5 driverless operation. V2X networking and connectivity enables sensor fusion across vehicles and vehicle infrastructure and therefore additional use cases like remote sensor and collective perception get applicable. Crucial use cases are the ones which ensure and extend active safety for vehicles beyond line-of-sight with cooperative adaptive cruise control and intersection collision warning. And for example, V2X networking and connectivity is a key enabler of the intelligent transportation systems (ITS) ecosystem. And there are finally the most likely and quite soon feasible business model driven use cases like vehicle-, mobility-, freight-as a service, and smart cities. But there is still an ongoing debate between all V2X ecosystem stakeholders about the business practicality of V2X networking and connectivity.

References

ISO 17515-1. (2015). Intelligent transport systems – Communications access for land mobiles (CALM) – Evolved Universal Terrestrial Radio Access Network (E-UTRAN) – Part 1: General usage. ISO.

3GPP TR 22.885 (2015). Study on LTE support for Vehicle to Everything (V2X) services. 3GPP.

3GPP TR 22.886 (2015). Study on enhancement of 3GPP Support for 5G V2X Services. 3GPP.

3GPP TR 22.891 (2015). Feasibility Study on New Services and Markets Technology Enablers. 3GPP.

3GPP TS 22.185 (2017). LTE Service requirements for V2X services; Stage 1. ETSI.

5GAA-WG1 T-170063 (April 2017). Use Cases Descriptions, Requirements and KPIs. 5GAA.

5G Americas (April 2017). White Paper on 5G Spectrum Recommendations.

5GCAR 5G Communications Automotive Research and innovation https://5gcar.eu/

5G PPP (2015). *5G Automotive Vision*. 5G PPP.

ABI research: V2X and Cooperative Mobility Use Cases. Q2 (2017)

Continental. (2017, July 28). *Holistic connectivity*. Retrieved from http://holistic-connectivity.com/

Daniel Watzenig, Martin Horn, Automated Driving – Safer and More Efficient Future Driving. *Springer International Publishing* Switzerland (2017).

Dieter Schramm, Manfred Hiller, Roberto Bardini, Vehicle dynamics. Springer-Verlag Berlin Heidelberg (2014)

DOT HS 811 492A (September 2011). Vehicle Safety Communications – Applications (VSC-A) – Final Report. DOT.

Dr. Meng Lu, Evaluation of Intelligent Road Transport Systems. The Institution of Engineering and Technology (2016)

ETSI. (June 2009). Intelligent Transport Systems (ITS); Vehicular Communications; Basic Set of Applications; Definitions. ETSI ITS Specification TR 102 638 V1.1.1. ETSI.

Farradyne. (2005). *Vehicle infrastructure integration (VII) architecture and functional requirements*. U.S. Department of Transportation Federal Highway Administration.

Gerrit Meixner, Christian Müller, Automotive User Interfaces – Creating Interactive Experiences in the Car. *Springer International Publishing AG* (2017).

Huawei. (2016). Communications networks for connected cars. Huawei.

ISO 13111-1 (2017). Intelligent transport systems (ITS) – The use of personal ITS station to support ITS service provision for travellers – Part 1: General information and use case definitions. ISO.

ISO/CD TR 17427 (2013). Intelligent transport systems – Cooperative ITS – Part 13: Use case test cases. ISO.

ISO/TR 13184-1 (2013). Intelligent transport systems (ITS) – Guidance protocol via personal ITS station for advisory safety systems – Part 1: General information and use case definitions. ISO.

ISO/TR 13185-1 (2012). Intelligent transport systems – Vehicle interface for provisioning and support of ITS services – Part 1: General information and use case definitions. ISO.

ISO/TR 17185-3 (2015). Intelligent transport systems – Public transport user information – Part 3: Use cases for journey planning systems and their interoperation. ISO.

ISO/TS 16460 (2016). Intelligent transport systems – Communications access for land mobiles (CALM) – Communication protocol messages for global usage. ISO.

Markus Maurer, J. Christian Gerdes, Barbara Lenz, Hermann Winner: Autonomous Driving – Technical, Legal and Social Aspects. *Springer Open Access* (2015).

Michael Huelsen, Knowledge-based driver assistance systems. Springer Fachmedien Wiesbaden (2014)

NGMN. (September 2016). Perspectives on Vertical Industries and Implications for 5G, Version 2. NGMN.

SAE J2735, Dedicated Short-Range Communications (DSRC) Message Set Dictionary.

Qualcomm. (2017, August 3). *Accelerating C-V2X toward 5G for autonomous driving*. Retrieved from
https://www.qualcomm.com/news/onq/2017/02/24/accelerating-c-v2x-toward-5g-
autonomous-driving

Use cases for autonomous driving according to Markus Maurer, J. Christian Gerdes, Barbara Lenz,
Hermann Winner: Autonomous Driving – Technical, Legal and Social Aspects. Springer Open
Access 2015 are interstate pilot using driver for extended availability, autonomous valet park-
ing, full automation using driver for extended availability and vehicle on demand.

Chapter 3
V2X Requirements, Standards, and Regulations

In this chapter, we discuss requirements derived from use cases for vehicular communications and related activities in standards developing organizations (SDOs) and regulation bodies. In this discussion, a number of unexpected conclusions will come up—in particular, we discuss why vehicular communications can indeed support advanced solutions for assisted driving, but the real need for it will only occur in a later stage, when fully autonomous vehicles are deployed in large scale numbers. Assisted driving relies on a number of sensors and vehicular networking and connectivity play the role of an additional sensor even though its characteristics are different to traditional video camera, sonar, radar, LiDAR and related components. If vehicular communications links are available, they can enrich the sensor data and contribute to the overall decision making—in case of absence, however, other sensors will take over. Vehicular communications are not essential. Actually, it cannot play an essential role in the first place, since the availability of wireless communications links cannot be guaranteed to be available any-time, any-where without down-time.

We differentiate between in-vehicle and external networking and connectivity. In-vehicle networking and connectivity is currently comprised of smartphone integration (VehiclePlay, AndroidAuto, MirrorLink, SmartDeviceLink) with 3GPP and Wi-Fi, vehicle access with BLE and NFC, smartphone and audio video integration with Bluetooth, hotspot provision with Wi-Fi and sensor access with Bluetooth and Wi-Fi. External vehicle networking and connectivity have these major options which are DSRC and cellular 4G and the upcoming 5G. Wi-Fi and BLE play a minor role here. All these networking and connectivity options make the vehicle one of the most heterogeneous connected system platforms in the ecosystem. Taking this into account, automated and autonomous vehicles create additionally gigantic challenges for wireless access in terms of number of connected vehicles, magnitude of data to be up- and downloaded (100s of GB per day) and spectrum bandwidth needed.

These unique challenges are first, the presence of multiple radios for connected vehicles with issues like interference, interoperability and required update options due to the different technology life cycles of radio modems and vehicles. Second, there are mobility challenges due to intermittent connectivity and extended velocity ranges. Third, we see requirements steadily increasing due to the continuously changing environment and vehicle surroundings with a very dynamic topology, infrastructure constraints and uncertainties. Fourth, there are the safety, security and reliability needs with low latency, extremely low BER and BLER, tamper proof and data privacy. And finally, we expect answers to be given for the challenge of network congestion avoiding any loss of connectivity.

DOI 10.1515/9781501507243-003

The situation is very different for today's use cases with no automation or significant driver assistance from the future automated and autonomous driving use cases. Autonomous driving indeed requires a closer interaction between neighboring vehicles, joint decision making related to optimum vehicle speed, platooning behavior, etc. Other sensors cannot offer alternative solutions, a wireless data exchange between vehicles is required. Furthermore, the required capacity of wireless data exchange will be dramatically different for the autonomous driving case compared to the first generation exchange of simple safety messages. As is exhibited in Figure 3.1, thousands of Gigabytes are collected by a vehicle each day (versus 1.5GB for the average internet user per day) and at least a portion of the data are conveyed to fellow vehicles in the neighborhood or to remote data centers for further processing. Each vehicle generates about as much data as about 3,000 people – and a million automated and autonomous driving vehicles will generate 3 billion people's worth of data. The challenges for suitable wireless system designs are tremendous!

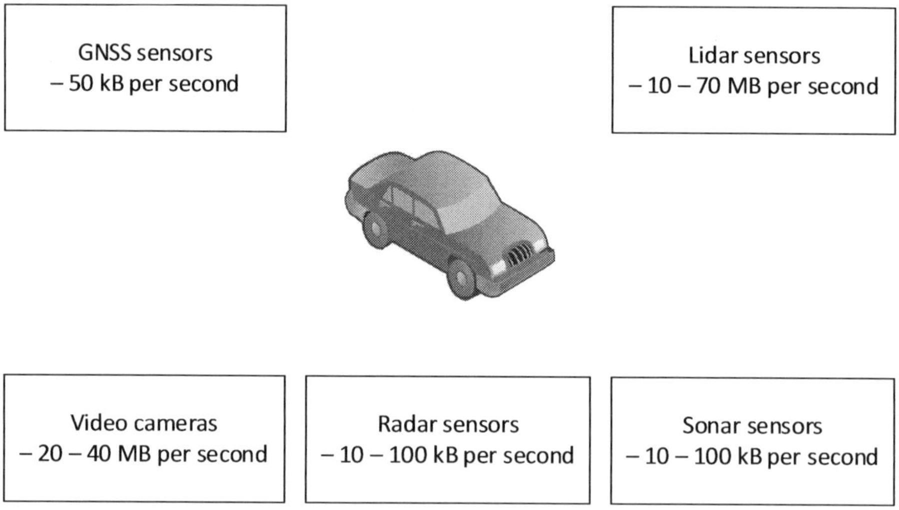

GNSS sensors	Lidar sensors
− 50 kB per second	− 10 − 70 MB per second

Video cameras	Radar sensors	Sonar sensors
− 20 − 40 MB per second	− 10 − 100 kB per second	− 10 − 100 kB per second

Figure 3.1: A vehicle creates thousands of Gigabytes per day

The importance of vehicular communications technology will thus increase over time and eventually become an essential part of the overall traffic management ecosystem. In the following sections, we will take a closer look at requirements of key use cases followed by a discussion of related activities in standards and regulation bodies. The latter are required to ensure an interoperable ecosystem enabling wireless interactions in a rich multi-vendor environment.

Autonomous or automated vehicles are at the top on the hype cycle. But the path to fully autonomous and automated driving is bumpy and full of surprises for vehicle makers, politicians, regulators and computing and communications engineers. We may end up in a fundamentally different vehicular future which is quickly approaching. Many vehicle ecosystem stakeholders have already got a fleet of self-driving vehicles populating the roads in Germany, United States, Japan and China since January 2017. And we anticipate more self-driving vehicles hitting the roads changing from tests to unrestricted operation in the coming years. The V2X communications systems behind the scenes, depend on a fast, reliable network that's capable of transmitting data and receiving from any kind of nodes.

The present-day test vehicles help us to acquire data to feed artificial intelligence and deep learning algorithms. Nevertheless, when we spot a self-driving vehicle of Audi and BMW on the highway on test today, this astonishing vehicle with no driver at the steering wheel is perceived as a single item. It's obvious to us that this cool vehicle has apparently nothing to do with the driving practice of anyone else on the road. However, this perception of singularity will change as more autonomous or automated vehicles appear on the road. Networking and connectivity are crucial when every unique autonomous or automated vehicle in fact become part of a very complex ecosystem and get linked to other vehicles, to road-side units, the infrastructure and finally data centers in the cloud. The more vehicles on the road, the more a high-performance, pervasive, reliable, and safe and secure networking and connectivity infrastructure is mandatory to unfold the real potential of autonomous and automated vehicles to the ecosystem. Some of these requirements are currently input to 3GPP 5G network standardization work, which is expected to support trials being already underway and to finally deliver in 2020.

3.1 Requirements

Before deriving technical networking and connectivity requirements from key use cases, we again bring up the evolution of automation in future vehicles. Table 3.1 refers to the generally acknowledged list of upcoming automation levels for automated and autonomous driving from SAE, starting with today's vehicles, which typically offer either no automation or driver assistance; these will evolve toward full automation of all driving modes in the future.

Table 3.1: System and driver roles in automated and autonomous driving levels as of SAE

SAE level	Execution of steering & acceleration/ deceleration	Monitoring of driving environment	Fall-back performance of dynamic driving task	System capability (automated driving modes)
0 (No automation)	Human driver	Human driver	Human driver	N/A
1 (Driver assistance)	Human driver & system	Human driver	Human driver	Some driving modes
2 (Partial automation)	System	Human driver	Human driver	Some driving modes
3 (Conditional auto-mation)	System	System	Human driver	Some driving modes
4 (High automation)	System	System	System	Some driving modes
5 (Full automation)	System	System	System	All driving modes

In the immediate future, assistance systems will support the driver with safety, environmental, and collective perception information. These will contribute to the objective to substantially increase road safety and to reduce the level of accidents overall. With the capacity of wireless data links increasing (enabled by the convergence toward fifth-generation cellular communications systems), drivers will furthermore be supported by high-quality and high-definition content for in-vehicle infotainment; related technical requirements include high throughput downlink (DL) features. With a future availability of high throughput uplink (UL), further services will become available such as HD map downloading in real-time. Finally, over-the-air updates (OTA) help to address security vulnerabilities that may occur over the lifetime of a vehicle (typically ten years or more), and to add new features to existing in-vehicle systems as vehicular communications standards evolve.

The implementation of V2X networking and connectivity solutions is already ongoing in use cases like emergency call, telematics, voice communications, internet access and infotainment which are part of SAE level 0, but no automation. In the next phase, V2V networking and connectivity shall enable many new cooperative use cases aimed at reducing road hazard situations as well as enhancing traffic efficiency and individual driving comfort, supporting an evolved ADAS which corresponds to SAE level 1. SAE levels 2 to 5 are supported by V2X networking and connectivity as a sensor extension. Additionally, new business models are going to drive V2X toward this phase. When the focus of V2X use cases reaches SAE level 1 use cases and evolves in addition, the performance of V2X networking and connectivity has to satisfy the performance requirements of both, non-safety and safety use cases.

V2X networking and connectivity use cases have a need of a periodic message exchange over the wireless interface and the use case reliability is determined directly by the communications link quality. The wireless link quality is affected by the surrounding environment, the transmitter and receiver mobility patterns, location, network load and many other parameters. Reviewing and estimating the main influencing factors on the communications link quality in V2X networking and connectivity use cases are necessary to decide upon the required frequency spectrum, to evaluate the various communications technologies and system architectures to fulfill requested use case reliability. Studying the limits of V2X networking and connectivity technology options is necessary to understand when V2X networking and connectivity use cases cannot be supported efficiently anymore and what the constraints are.

Currently there are metric proposals of communications link quality for V2X networking and connectivity use cases. Requirements and metrics specified by communications and computing stakeholders are typically derived from classical wireless network performance metrics assuming a static infrastructure and a certain mobility pattern of mobile devices. And requirements and metrics stated by vehicle industry stakeholders pay attention to specific traffic participant mobility patterns, reliability and suitability for both safety and non-safety applications. Examples are the DSRC classical network metrics such as packet delivery ratio (PDR) and received signal strength indication (RSSI). Both, PDR and RSSI do not provide a relationship between use case reliability requirements and communications performance. In addition, there is the huge variety of V2X networking and connectivity use cases that perform differently for each communications technology due to their very specific communication reliability requirements. For example, timing periods of messages are critical for ADAS use cases such as lane change assistant or emergency brake light warning, since data about other vehicles, pedestrians etc. change continuously. Otherwise dissemination distances are critical for V2X use cases like traffic monitoring, weather data or navigation since data are valid for longer period of times.

The evaluation metric gives an insight where the use case reliability starts getting out of required values. The metric shall work for simulations as well as operational field trials data, dealing with timing as well as distance properties. Metrics to start with and to evaluate communications protocol quality are latency, coverage and message loss ratio measured as packet delivery ratio (PDR, source simTD). PDR is defined as the percentage of the number of successfully received messages divided by the number of all messages sent and is used as a function of distance. PDR comprises typical impairments of wireless communications like fading and access collisions. PDR is zero at the maximum communications distance. The number of messages required to get one message certainly received is derived from PDR as an inverse function for an optimistic value or takes the success probability into account for a more realistic value. If PDR and latency get combined we get the recently proposed maximum consecutive CAM period (CCP) to measure the reception and time.

V2X networking and connectivity technology candidates to support intelligent transport systems (ITS) have gotten a lot of attention for many years. It's all V2X ecosystem stakeholders' task to evaluate the V2X networking and connectivity use cases regarding spectrum demand, communications protocols and safety and security impact. Due to the exceptional challenges and requirements in particular for communication, safety and security, it is necessary to analyze and determine ideally the capabilities of these technology candidates very early in the system design. But it's very difficult and expensive to evaluate the huge amount of use cases in the real world. We see two solutions addressing this issue, first these are simulation environments as close as necessary to the real world and second operational field trials and tests.

The challenge for simulation environments is to integrate the two different worlds of vehicular traffic simulation tools with wireless network simulation tools and geographic 3D road maps using an integrated flexible simulation framework to accurately simulate V2X communications with realistic mobility, accurate propagation models and precise communications models. Additionally, interfaces are needed to plug-in different communications, safety and security protocols. Traffic simulators like VISSIM or SUMO provide the right mobility models but lack support for communication protocols. Network simulators like NS-3 or OMNET++ are designed for evaluation of general purpose network protocols and lack support of exclusive mobility and precise propagation models required for V2X networking and connectivity use cases. Wireless radio wave propagation simulators as of 3GPP provide propagation models. The VSimRTI simulator is capable of integrating several simulators from different domains and allows the use of realistic movement dynamics for V2X traffic stakeholders in relationship with network and wireless propagation characteristics therefore. On the other hand, none of the simulation tools currently provide a complete view on the impact of V2X networking and connectivity on spectrum demand, communications performance, and infrastructure and business models.

Operational field trials are the alternative option to evaluate V2X networking and connectivity technology candidates. But these field trials become quickly expensive since they require a well set up infrastructure including vehicles and other traffic stakeholders. Trials in real traffic environments are not an option since it's very challenging to arrange the traffic stakeholders in the exact way the use cases demand. In both cases, it's very important to sample the event context together with timing and location data.

We will give a short overview about self-driving vehicle operational field trials done by vehicle ecosystem stakeholders up to now and planned for 2017. Uber did pilot tests for its self-driving vehicles in Pittsburgh, San Francisco and in Arizona in 2016, but suspended them in 2017. Tesla started installing tests for full autonomy in October 2016 and planned a full self-driving test from Los Angeles to New York before the end of 2017. Google's Waymo has been testing a fleet of roughly 60 self-driving vehicles in Mountain View, Phoenix, Austin, and Kirkland in 2016. Waymo planned to test a fleet of roughly 100 self-driving Chrysler minivans on public roads in 2017.

Baidu teamed up with China-based vehicle manufacturer BAIC to test Level 3 self-driving vehicles on public roads in 2017. Volvo planned to test 100 self-driving vehicles on public roads in Gothenburg and in China in 2017. Ford begins testing its self-driving vehicles with a fleet of up to 100 vehicles in Europe in 2017. GM also planned to test its self-driving vehicles in Detroit, San Francisco and Scottsdale in 2017. BMW is testing a fleet of self-driving vehicles in Munich in 2017.

3.1.1 Sensors

With the evolution from driver assisted driving to high and full automation, many vehicle sensors are already supporting ADAS today and play a key role for that reason. Going forward, the vehicle on-board sensor capabilities used for driver assistance and active safety systems today get extended in terms of performance and deliver increasing amounts of data supporting various light and weather conditions and ranges. Moreover, these sensors provide a host vehicle with surrounding traffic stakeholders' data such as speed, acceleration, location, etc. Sharing these sensor data leads to more efficient vehicle traffic and avoids in some cases the implementation of high-cost sensors in any vehicle and vehicle infrastructure. Commonly used sensors for automated and autonomous driving vehicles are illustrated in Figure 3.2. Several thousands of Gigabytes of data will be generated by each vehicle every day. Part of these data are transferred to data centers in the cloud in order to feed artificial intelligence and deep learning algorithms. Corresponding decision making features will thus constantly be updated and fed with the latest data.

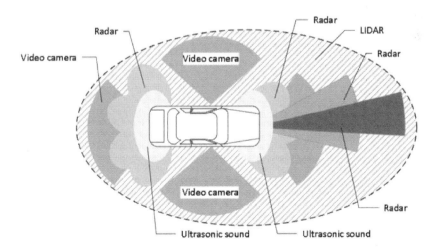

Figure 3.2: Vehicle sensors for automated and autonomous driving

Typical examples of V2X networking and connectivity use cases, exploiting sensors as vehicle extensions, are emergency electronical brake light (EEBL), forward collision warning (FCW) intersection collision warning (ICW), stationary vehicle warning (SVW), and traffic jam warning (TJW) or pre-crash warning (PCW).

3.1.2 Communications

Connected vehicle technology is rapidly evolving to include many V2X networking and connectivity technologies for vehicle-to-vehicle (V2V), vehicle-to-infrastructure (V2I), vehicle-to-device (V2D), and vehicle-to-pedestrian (V2P) messaging and communications. In addition, the integration of V2X networking and connectivity, leveraging for example LTE, 5G and DSRC, with non-vehicle use cases implemented through various vehicle-to-home (V2H), vehicle-to-device (V2D), and vehicle-to-grid (V2G) solutions creates plentiful of new business opportunities in the area of mobility as a service.

SAE and ETSI ITS outline use cases for V2X networking and connectivity implementing DSRC for data exchange between a host vehicle and remote vehicles and between a host vehicle and the roadside, to address safety, mobility, and environmental system needs. The objective is to send vehicle non-safety and safety messages with a required reliability and needed delay. The communications is one-to-many, local, or geo-significant supported by a vehicular communications network which is set up ad-hoc, highly mobile, and with sometimes a large numbers of communications nodes. DSRC supports messaging for non-safety use cases like intelligent transport systems (ITS) applications such as electronic toll collection (ETC), exchange of speed limits, points of interest or digital map updates as well as safety-critical ones for example like collision warnings, emergency breaking lights, traffic condition warnings or lane assist. The V2X networking and connectivity requirements are given in the Tables 3.2, 3.3 and 3.4.

Table 3.2: V2X networking and connectivity requirements as of DSRC

Feature	Intersection collision avoidance	Cooperative collision warning	Toll collection	Infotainment
Message rate and size in Hz and bytes	10 and 100	10 and 100	NA and 100	NA and streaming
Latency in milliseconds	Up to 100	Up to 100	Up to 50	Up to 100
Communications range in meters	Up to 300	Up to 300	Up 15	Up to 100
Reliability in %	99.999	99.999	99	95

Table 3.3: Safety messages requirements

Use case	Message type	Communication mode	Message rate in Hz	Latency in milliseconds
Emergency electronic brake lights	DEN / V2X	Time limited, event-based periodic broadcast	10	100
Abnormal condition warning	DEN / V2X	Time limited, event-based periodic broadcast	1	100
Emergency vehicle warning	CAM / V2X	Periodic broadcast vehicle-mode dependent	10	100
Slow vehicle warning	CAM / V2X	Periodic broadcast vehicle-mode dependent	2	100
Motorcycle warning	CAM / V2X	Periodic broadcast	2	100
Vulnerable road user warning	CAM / VRU2X	Periodic broadcast	1	100
Wrong way driving warning	DEN / V2X	Time limited, event-based periodic broadcast	10	100
Stationary vehicle warning	DEN / V2X	Time limited, event-based periodic broadcast	10	100
Traffic condition warning	DEN / V2X	Time limited, event-based periodic broadcast	1	100
Signal violation warning	DEN / V2X	Time limited, event-based periodic broadcast	10	100
Roadwork warning	DEN / V2X	Time limited, event-based periodic broadcast	2	100
Overtaking vehicle warning	DEN / V2X	Time limited, event-based periodic broadcast	10	100
Lane change assistance	DEN / V2X	Time limited, event-based periodic broadcast	10	100
Pre-crash sensing warning	DEN / V2X	Time limited, event-based periodic broadcast	10	50
Cooperative glare-reduction	DEN / V2X	Time limited, event-based periodic broadcast	2	100

Table 3.4: Non-safety messages

Safety service	Use case	Message type	Communication mode	Minimum updates per second	Maximum latency in milliseconds
Traffic management	Speed Limits	I2V	Periodic broadcast	2	100
	Traffic lights optimal speed advisory	I2V	Periodic broadcast	1	100
	Intersection Management	I2V	Periodic broadcast	1	100
	Cooperative flexible lane change	I2V	Periodic broadcast	1	500
	Electronic toll collection	I2V	Periodic broadcast	1	500
Infotainment	Point of interest notification	I2V	Periodic broadcast	1	500
	Local electronic commerce	I2V, V2I	Duplex, internet access	1	500
	Media download	I2V	Duplex, internet access	1	500
	Map download and update	I2V	Duplex, internet access	1	500

3GPP defined V2X use cases for LTE for road safety and traffic flow management, for instance: forward collision or control loss warning, emergency stop warning, queue warning, warning to pedestrians against collision or cooperative adaptive cruise control. 5G NR extended these use cases by adding ones for automated and autonomous driving and infotainment. Examples are the vehicle as a mobile office, semi-autonomous and fully autonomous driving including cooperative driving, traffic and route management or services for commercial road users (e.g. truck platooning). The related communications requirements for LTE and 5G NR are summarized in Table 3.5.

Table 3.5: V2X networking and connectivity requirements as of 3GPP

Feature	LTE	5G NR
End-to-end latency in milliseconds	100	1, 5 and less than 10
Communication range in meters	Up to 320	Up to 1000
Absolute and relative speed in km/h	160 and 280	Up to 240 and up to 500
Message rate and size in Hz and bytes	10 and up to 1200	Up to 100 and 1600
Position accuracy in meters	Less than 15	Absolut less than 1, relative less than 0.1
Reliability in %	90	Up to 99.999

3.1.3 Dynamic high-definition maps

Dynamic high definition (HD) maps are a key requirement for today's driving and even more asked for in future automated and autonomous driving. Any incorrect representation of the road infrastructure can indeed lead to critical and possibly fatal situations. The challenge is that the current road network situation is in constant evolution. Local road work efforts, closed roads and similar changes may occur at any time and need to be made available to the automated and autonomous driving system. Local dynamic HD maps therefore need to be constantly updated through over-the-air wireless services. In practice, cloud-based mapping services are used in order to support connected ADAS and autonomous driving features. Extremely detailed timely maps down to the nearest centimeters, showing lanes and road hazards, among other things will thus be constantly available to the user.

The static basic map layer with digital cartographic and topological data is a precise, down to sub-lane level accuracy, representation of the road facilities including slope and curvature, static lane marking types and static roadside objects used already for telematics and localization. A semi-static map layer is put over, for instance for traffic signs, landmarks, traffic regulations, road work and weather forecasts. A semi-dynamic map layer is overlaid including data for example of road conditions, traffic congestion, accidents, local weather conditions and traffic signal phases. A highly dynamic map layer is used for providing, for instance, data of surrounding vehicles, pedestrians and road hazards. And an analytics layer overlay delivers tools to compare the data from vehicle sensors with the data that are on the map layers to analyse the differences and use them for example to update a common map database or maintain vehicle driving profiles.

The data gathering for dynamic HD maps happens with very different sampling times for each layer, around seconds for dynamic data, around minutes for semi-dynamic data, around hours for semi-static data and around weeks and months for static data. All data get integrated into a common database for the local dynamic HD map. Data sources are any stakeholder of the traffic ecosystem, from vehicles, pedestrians up to government organizations and agencies. A common sensor data ingestion interface specification defines the technical requirements of data sets for vehicles to transfer on-board sensor gathered data. The V2X networking and connectivity requirements are very diverse for the different local dynamic HD map layers.

The static basic map layer with digital cartographic and topological data merges digitized cartographic and topological maps with data collected by video camera or LiDAR equipped vehicles and other transmitted data about the driving path's geometry and stationary landmarks around each vehicle. The map-relevant data are analysed offline or in real-time in the vehicle and are sent to a common database. The basic map layer could get maintained by crowdsourcing real-time road data from vehicles. The data are then packed into small packages and sent to the cloud, the cloud server aggregates and reconciles the continuous stream of information producing a highly accurate and low time to reflect reality map. In case that a high capacity wireless link is not available, the system leverages video camera, radar, ultrasound and LiDAR data to create high-precision 3D base-maps. Then, the dynamic HD map is self-maintaining through a tile-based approach and the map's update is done later to reflect real-time road conditions changes. Since local dynamic HD maps are represented by multiple data tiles and layers, any updates to be conveyed to specific vehicles can be limited to the evolving data elements only and thus the over-the-air capacity requirements are less demanding. In this way, dynamic and semi-static map layer data of today are going to become the local basic map.

The use of V2X networking and connectivity for semi-dynamic map layer data may be advantageous for certain dedicated use cases like variable speed limit, variable traffic signs and signals, but needs to be aligned with traffic regulation and authorities. More useful is V2X communications as sensor extension for example for eco-driving use cases exploiting data of road conditions, traffic congestion, accidents, local weather conditions and traffic signal phases. The data for the highly dynamic map layer are too dynamic to take advantage of V2X networking and connectivity today due to their latency requirements and the challenges arising due to the parallel processing of control loops completely in the vehicle and control loops including V2X communications like remote control. In particular, safety use cases involving surrounding vehicles, pedestrians and road hazards are still out scope for quite some time.

Evolving toward SAE level 4 and 5, driving dynamic HD map updates shall minimize dynamic V2X communications, because the vehicle shall function independently of any external communications links. Both map data messaging and sensors add complexity and additional cost to the system. Data that are related to vehicle surrounding context and out of on-vehicle sensor range shall be communicated, when they do not duplicate data already available and provide updates to map layers in the common database. Currently the focus of map suppliers is on use cases like precise positioning, traffic light detection, speed limit compliance, automatic lane changing, automatic overtaking and collision avoidance.

3.1.4 Over-the-air updates

Connected vehicles are increasingly enabled to receive remote over-the-air (OTA) firmware and software updates from the supporting vehicle infrastructure. And vehicles transmit diagnostic and operational data from on-board systems and components and vice versa. By leveraging V2X networking and connectivity, vehicle manufacturers reduce significantly recall expense, improve cybersecurity response time, increase product quality and operational efficiency, and deliver post-sale vehicle performance and feature enhancements. These updates for software reliant systems and components, and data transmission power all use cases by offering opportunities for new services and business optimization efforts.

The challenges related to V2X networking and connectivity technology are exemplarily for many other vehicle system components. One challenge is to ensure that a radio communications component remains relevant over the entire lifetime of a vehicle, i.e. ten years and beyond. It is almost certain that a V2X framework feature-set will evolve within this period. Software reconfiguration will enable manufacturers to replace specific software and thus maintain related up-to-date feature-sets without requiring changes to the hardware. This approach reduces the overall cost for change since a vehicle does not need to be upgraded by an authorized dealer as it would be required for hardware changes, but the process is handled through over-the-air remote control. In extreme cases, for example for addressing platform vulnerabilities which may arise suddenly over the lifetime of a vehicle, there is even no time for recalling and manually updating millions of vehicles; rather, immediate action is required in order to ensure the safety of the passengers and other road users. Over-the-air software reconfiguration provides an efficient solution to deal with these issues.

Other challenges are regulatory driven rollback and recovery of software based features for a huge amount of vehicles in the field. This includes a standards-based certification, authentication, and encryption roll out process based upon V2X networking and connectivity. Dynamic data collection and upgradable analytics are required to support the most efficient network and to access technology selection, possibly data caching, and data transfer.

3.1.5 In-vehicle Infotainment

High-definition content is expected to substantially support the driver through in-vehicle infotainment and any automated decision-making units. Typical examples for such content are road hazard awareness, road and traffic surveillance, smart traffic lights, virtual and augmented reality (VR, AR), infotainment, and productivity. For road hazard awareness, autonomous driving vehicles' cameras gather video data on road hazards and create crowd-sourced video data. For example, an automated version of a Waze-like community with rich visualization and possible immersion for passengers creates data for traffic awareness and road hazard awareness. For roads and traffic surveillance, road cameras monitor traffic and any road hazards and send this data to the cloud or the edge cloud. Partial processing can take place in the road cameras. Cloud or edge cloud services analyze the video data and can serve multiple purposes such as traffic service, city security service, and so on.

For smart traffic lights, cameras co-located with traffic lights monitor the traffic light intersection, vehicle' density, and pedestrians, and analyze the video data to control red or green traffic light duration. Each traffic light sends video data about its intersection to an edge cloud for the synchronization of multiple traffic lights on the road. For virtual and augmented reality, head-mounted devices allow passengers to have an immersive environment of the vehicle surroundings.

Infotainment services allow passengers to get access to their social media, movie-streaming services, online gaming, view vehicle data, or connect various other apps and handheld devices to displays in the vehicle. For productivity, office services allow passengers to have video conferencing sessions in vehicles.

In order to update the data to be provided for each vehicle, big data processing in the data center is required based on distributed data obtained from the field. Selected sensor data of vehicles are transferred to suitable cloud locations. Then it is required to feed and exploit learning models that are typically initially preconfigured during the production phase. Thus, the provided data continuously improves the automated and autonomous driving functions during the deployment of autonomous vehicles. For example, if an autonomous vehicle visualizes a scene that cannot be interpreted, data of the scene are sent to the data center for more learning and get used to update the machine learning models in the vehicle. Some typical examples for this kind of related data handling are the data copied from in-vehicle storage to the data center and data parsing to update data formats to prepare for processing, the data analytics to identify objects, scenes, and events, the data assessment to give recommendations on the data usage and coverage, and the hierarchal data storage or the search data or retrieval.

The infotainment evolution might even replace the vehicle-to-cloud approach with a visual cloud implementation. Such a visual cloud comprised of cloud computing architectures, cloud scale processing and storage, and ubiquitous V2X networking and connectivity between vehicles and the vehicle infrastructure, network edge vehicles and cloud data centers. Automated and autonomous driving challenges the in-vehicle only approach—the local computing power equipped in a self-driving vehicle is not enough due to space, heat dissipation, and the cost of executing the heuristics of artificial intelligence needed to provide such levels of autonomy. This drives the need to execute the complex and necessary intelligence in the cloud.

The development of automated and autonomous driving vehicles requires the processing, storage and exchange of huge amounts of generated sensor data from the vehicle including video data, which challenges not only cellular wireless networks or DSRC systems but also the upcoming 6G NR from a capacity and cost perspective. An efficient V2X networking and connectivity technology will help to evolve the implementation of visual computing applications that rely on cloud architectures, cloud scale computing or storage, and ubiquitous broadband interconnectivity between in-vehicle platforms, edge network platforms, and cloud data centers.

In this context, current limitations of vehicle-embedded systems need to be addressed. These include the sophisticated system design, increasing development cost and time, and the necessity to achieve high reliability and safety combined with large-scale and complicated software. The rising number of ECUs with increasing cost and space constraints for ECUs in the vehicle, and the complicated network architecture, lead to even more design complexity. A way forward is platform-based development including hardware, software, and the network including software. The vehicle infotainment system gets described as a set of software components which get connected via a virtual function bus. Then the system is mapped to ECUs and connected with the in-vehicle network.

3.2 V2X networking and connectivity standards

We start with a communications-centric look at autonomous and automated driving. The US Society of Automotive Engineers (SAE) adopted its standard J3016 targeting autonomous and automated vehicles in January 2014. The role allocation between human drivers and automated driving systems is specified by the six levels of driving automation. In Level 0 is pretty basic without automation. The driver controls it all, steering, braking, throttle, power. It's what you've been doing all along. Feet off and driver in constant control.

Level 1 is driver assisted, this semi-autonomous level means that most functions are still controlled by a driver, but some, like braking, can be done automatically by the vehicle. Feet off and driver constantly monitors and gets longitudinal or lateral assistance. There are at least two functions automated at level 2 with partial automation, like cruise control and lane entering. This means that the driver is disengaged from physically operating the vehicle by having his or her hands off the steering wheel and foot off the pedal at the same time. The driver must be ready to take control of the vehicle, however. Feet off and partly hands off and driver constantly monitors and gets longitudinal or lateral assistance.

Level 3, with conditional automation is where drivers are still necessary, but are able to completely shift safety-critical functions to the vehicle, under certain traffic or environmental conditions. It means that the driver is still present, but is not required to monitor the situation in the same way it does for the previous levels. Feet and hands off, partly eyes off, but the driver is in position to retake control; it isn't required to constantly monitor. In level 4 we see high automation where vehicles are designed to perform all safety-critical functions and monitor roadway conditions for an entire trip. Feet, hands and eyes are off the controls and no driver is required in certain scenarios. And level 5 with full automation refers to fully-autonomous vehicles that do not have any option for human driving—no steering wheel or controls. Feet, hands, eyes and mind off, no driver is required.

V2X networking and connectivity involves communication types resembling local- and wide area wireless communication. Specifically designed protocols as of IEEE WAVE, ETSI ITS-G5, ARIB T109 and IEEE 802.11p provide connectivity between V2X components like vehicles, roadside units (RSU) and pedestrians. Vehicle manufacturers rolled out V2X systems with optional communications packages with a 10 MHz spectrum at 760 MHz and additionally 90 MHz between 5.775 to 5.845 MHz in Japan in 2016. The U.S. federal government's V2V mandate proceeded with 75 MHz between 5.850 and 5.925 MHz for in 2016. There are 70 MHz between 5.855 and 5.925 MHz and 1 GHz between 63 and 64 GHz for V2V and V2R communications in Europe.

Currently there are two main technologies competing to be used for V2X networking and connectivity: IEEE 802.11p based DSRC (dedicated short range communications) and 3GPP LTE C-V2X. As it is illustrated below, IEEE 802 and 3GPP focus on the definition of lower layers only while the upper layers rely on standards by IEEE 1609, the Society of Automotive Engineers (SAE) and the Internet Engineering Task Force (IETF).

We start by taking a quick look at the status of the SAE V2X communications standards, which focus primarily on DSRC. An updated version of the SAE J2735 standard, which specifies a message set, data frames and elements specifically for 5.9 GHz DSRC/WAVE, was published in March 2016. SAE J2945/0 is in a ballot state that is going to create systems engineering principles for the J2945 suite of standards. SAE J2945/1, published in March 2016, defines the minimum performance requirements necessary to provide interoperability between on-board units for V2V safety systems. SAE J2945/2 outlines the DSRC requirements for V2V safety awareness and is currently a work in progress. SAE J2945/6 specifies the data exchange and performance requirements for coordinated and cooperative adaptive cruise control and platooning and is a work in progress. SAE J2945/9 will define vulnerable road user use cases for cooperative ITS and is in the ballot state. And SAE J3061 providing cyber-security guidance for automotive systems was issued in January 2016.

IEEE 802.11p work had been initiated in November 2004 and was finally completed in July 2010, with the approved *"Amendment 6: Wireless Access in Vehicular Environments"* to IEEE 802.11. Since then, the IEEE 802.11p based DSRC system has been extensively tested and validated. 3GPP completed the development of first generation vehicular services in the 3GPP Release 14 in 2017. Concerning the choice of a suitable system, the industry is currently in a dilemma: IEEE 802.11p based DSRC is currently available, but no further evolution is expected, so the system is likely to become obsolete in the long run. 3GPP LTE C-V2X, on the other hand, is going to evolve further toward 5G and thus is expected to take over the market in the long term; for the immediate usage, however, no products are available and thus manufacturers have little choice but to implement IEEE 802.11p based DSRC for the time being and foresee a migration towards 3GPP LTE C-V2X in the mid/long term. To make the problem even more challenging, both systems are fully incompatible with each other. Moreover, the whole deployment and challenge usage philosophies are built on substantially different principles: While IEEE 802.11p based DSRC applies distributed channel access based on IEEE 802.11's Carrier Sense Multiple Access (CSMA), LTE C-V2X builds on the centralized network architecture and scheduled channel access as it is known from commercial cellular LTE networks.

As illustrated below, the entire IEEE 802.11p based DSRC system builds on the following standards. The physical (PHY) layer and medium access control (MAC) layer use IEEE 802.11p with 802.11, which is the base standard of products marketed as Wi-Fi extensions required to support intelligent transportation systems (ITS) applications. The medium access control (MAC) layer provides extensions of IEEE 1609.4 including parameters for priority access, channel switching and routing, management services, and primitives designed for multi-channel operations. The logical link control (LLC) as the upper part of the data link of the OSI Model is defined by IEEE 802.2. LLC is a software component that provides a uniform interface to the user of the data link service, usually the network layer. Towards the lower layer, LLC uses the services of the Media Access Control. The network and transport layer, called WAVE short message protocol (WSMP), is the IEEE 1609.3 standard for wireless access in vehicular environments with (WAVE) providing the networking services. Layer 3 and layer 4 of the open system interconnect (OSI) model are implemented with the Internet Protocol (IP), user datagram protocol (UDP) and transmission control protocol (TCP) elements of the internet model. Management and data services within WAVE devices are provided as well. IETF RFC 2460 specifies version 6 of the internet protocol (IPv6), also sometimes referred to as IP Next Generation or IPng. SAE J2735 specifies a message set, and its data frames and data elements specifically for use by applications intended to utilize the 5.9 GHz Dedicated Short Range Communications for Wireless Access in Vehicular Environments, communications systems. SAE J2945 serves as the parent document for the J2945/x family of standards. It contains cross-cutting material which applies to the other J2945 standards, including guidance for the use of Systems Engineering (SE) and generic DSRC interface requirements content. The scope of the J2945 system environment is the information exchange between a host vehicle and remote vehicles and between a host vehicle and the roadside, to address safety, mobility, and environmental system needs.

IETF RFC 793 defines the Transmission Control Protocol (TCP), IETF RFC 768 defines the User Datagram Protocol (UDP). And IEEE 1609.2 defines secure message formats and processing for use by Wireless Access in Vehicular Environments (WAVE) devices, including methods to secure WAVE management messages and methods to secure application messages. It also describes administrative functions necessary to support the core security functions.

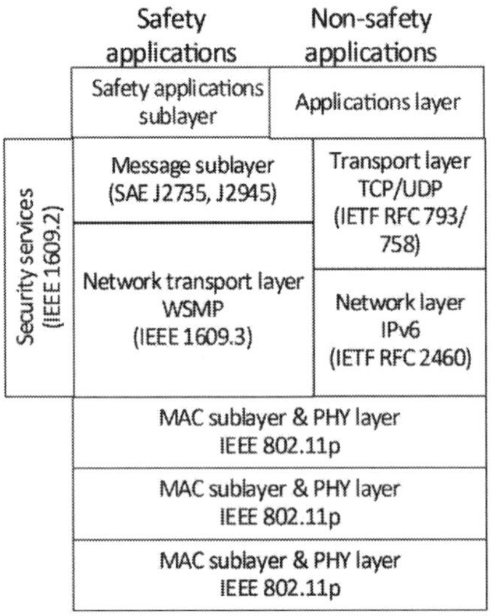

Figure 3.3: IEEE 802.11p based DSRC protocol stack

While the DSRC standard has already been deployed in several trials, especially in North America, only in September 2016, 3GPP issued the first release of its LTE V2X standard and will further work on an evolution in 3GPP Release 15 and possibly beyond. 3GPP builds on the cellular architecture indicated in Figure 3.4 and focuses on the definition of the lower layers which are in the explicit scope of 3GPP.

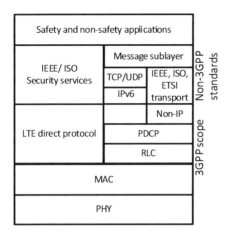

Figure 3.4: LTE C-V2X protocol stack

The upper layers, marked as "non-3GPP standards" in Figure 3.4, are exploiting the same standards as IEEE 802.11p based DSRC—see list above. 3GPP, as part of the expansion of the LTE platform to new services, started to work on developing functionality to provide enhancements specific for vehicular communications both in terms of direct communications between vehicles, and vehicles to pedestrian/infrastructure, and cellular communications with networks. The work has been organized in V2V (RP-161272) and V2X (RP-161298) Working Items (WIs). The V2V WI specifies the details of vehicle to vehicle communications, and has provided enhancements to earlier work on Device-to-Device (D2D) Proximity Services (ProSe) and the PC5 interface while the V2X WI is focused on additional aspects of vehicular communications leveraging the cellular infrastructure. The corresponding functionalities are included in 3GPP's Release 14 and will further be evolved in subsequent releases.

The upper IEEE and 3GPP standards are further complemented by activities in ETSI's Intelligent Transport System (ITS) Technical Committee: ETSI TC ITS develops global standards for Co-operative ITS, with a focus on vehicle-to-vehicle and vehicle-to-roadside fleet and communications. Applications include road safety, traffic control, fleet and freight management and location-based services, providing driver assistance and hazard warnings and supporting emergency services.

ETSI TC ITS develops standards related to the overall communications architecture, management (including e.g. Decentralized Congestion Control), security as well as the related access layer agnostic protocols: the physical layer (e.g. with ITS-G5), Network Layer, Transport Layer (e.g. with the GeoNetworking protocol), Facility Layer, (e.g. with the definition of facility services such as Cooperative Awareness—CA, Decentralized Environmental Notification—DEN and Cooperative Perception—CP, used by the ITS applications). Other addressed topics include, among other things, platooning, specifications to protect vulnerable road users such as cyclists and motor cycle riders, specifications for Cooperative Adaptive Cruise Control as well as multichannel operation. ETSI TC ITS also develops conformance test specifications which are crucial for the commercial deployment of the technology for all the above standardization activities.

Key ETSI ITS activities are summarized in the sequel at www.etsi.org/technologies-clusters/technologies/automotive-intelligent-transport. Cross layer Decentralized Congestion Control (DCC) provides stability in the ad-hoc network by providing resource management when there are a high number of C-ITS messages in order to avoid interference and degradation of C-ITS applications. C-ITS and safety driving applications depend upon reliable and trustworthy data transmitted by other vehicles and the infrastructure. In this context, standardized solutions for security and privacy are paramount and this will be based on the design and implementation of a security management infrastructure for cooperative-ITS. ETSI TC ITS develops standards defining the security framework for cooperative ITS including a PKI. This security framework will support PKI trust model requirements from the EU C-ITS deployment platform and bring privacy protection mechanisms for users and drivers, e.g. using pseudonym certificates and regularly changing pseudonym IDs in ITS G5 communications.

Many ITS applications require the dissemination of information with a rapid and direct communications, which can be achieved by ad hoc networking. GeoNetworking (GN) is a network layer protocol for mobile ad hoc communications without the need for a coordinating infrastructure based on wireless technology, such as ITS-G5. It utilizes geographical positions for dissemination of information and transport of data packets. It offers communications over multiple wireless hops, where nodes in the network forward data packets on behalf of each other to extend the communications range.

ETSI TC ITS develops and maintains important services to be used by ITS applications. These services include but are not limited to cooperative awareness (CA) to create and maintain awareness of ITS-Ss and to support cooperative performance of vehicles using the road network, decentralized environmental notification (DEN) to alert road users of a detected event using ITS communications technologies, cooperative perception (CP) complementing the CA service to specify how an ITS-S can inform other ITS-Ss about the position, dynamics and attributes of detected neighboring road users and other objects and multimedia content dissemination (MCD) to control the dissemination of information using ITS communications technologies.

V2X networking and connectivity must support functional safety when the technology gets implemented for non-safety use cases as well as for safety use cases. Safety is defined as the freedom from those conditions that can cause death, injury, occupational illness, or damage to or loss of equipment or property, or damage to an environment whereas reliability is the ability of a system or component to perform its required functions under stated conditions for a specified period of time. See Figure 3.5. A safe system is not necessarily reliable and vice versa. "Functional safety is the absence of malfunctioning behavior of E/E systems" (ISO 26262-1). The ISO 26262 functional safety standard got published in November 2011, offering a general approach for the development of E/E systems. The use cases regarding automated driving are not currently covered by the functional safety standard ISO26262. The next version of the standard will be published in 2018. ISO 26262 is a compilation of best practice for developing a safety-related system. In the standard, many development techniques of safe systems are listed and required depending on the safety level (ASIL). Reliability of hardware is calculated from the failure rate of each component of the system. Reliability of software is achieved within the software development process. Software that is developed with a process satisfying the requirements of the standard is considered to be reliable enough. ISO 26262 consists of 10 parts (Part 10 has not been published yet).

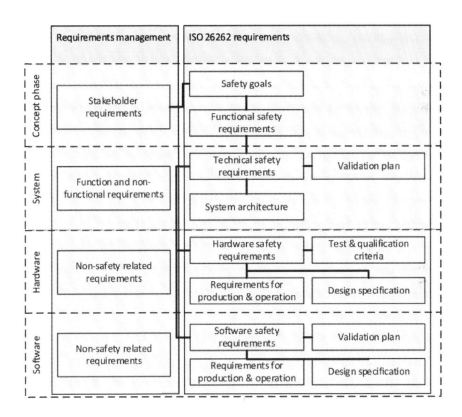

Figure 3.5: ISO safety requirements

3.3 V2X networking and connectivity regulation

As more and more fully automated, and ultimately autonomous, vehicles are tested on public roads in Europe, China, Japan and the United States, regulation has to be set up accordingly. Local national regulation authorities need to either adjust quickly existing rules or adapt new ones, Regulation is necessary to ensure that the automated or autonomous vehicles are in compliance with the developers', operators' and owners' expectations regarding safety, legal responsibility and privacy. Among the demanding regulatory questions to be answered are ethical ones, for example what are the differences between human and autonomous crash decision making and how liability is determined.

For instance, Germany's Federal Government adopted guidelines with the "Automated and Networked Driving Strategy - Leading Provider, Leading Market, Initiating Regular Operation" after a conversation between economy, science, association and administration stakeholders. The settled recommendations for research areas in the development of autonomous and automated driving concluded in the "Automated and Networked Driving" directive that provides support for upcoming solutions.

And the levels of autonomous or self-driving technology were defined in close similarity to SAE levels adopted by the United States' National Highway Traffic Safety Administration (NHTSA) in 2013. Level 0 is comprised of warnings only or automated secondary functions such as headlights or wipers. In level 1, the vehicle controls one or more safety-critical functions, but each function operates independently. The driver still maintains overall control. Level 2 combines two or more technologies from level 1, and both levels operate in coordination with each other. The driver still maintains overall control. Level 3 provides limited self-driving technology. The driver is able to hand off control of all safety-critical functions to the vehicle, and only occasional control by the driver will be required. Level 4 is a completely self-driving vehicle. The vehicle will control all safety-critical functions for the entire trip.

The challenges are among others to find solutions, such as for the safeguarding and approval of autonomous and highly automated assistance features. For instance, if self-learning artificial intelligence systems get applied, functional safety and cybersecurity over the entire life cycle of the vehicle will have to be taken into account. Or solutions are required to approve smart transport infrastructures with V2X networking and connectivity, which link the various traffic participants by means of cooperative, reliable, safe and secure vehicle platforms,

Besides enabling V2X networking and connectivity technologies, the laws amending the road traffic acts come into force, which provide basic rules for the interaction of the driver and the vehicle with autonomous or highly or fully automated driving functions. It regulates rights and duties of the vehicle operator during the autonomous or automated driving phase, as well as arising liability issues. This sets up the necessary legal certainty for the drivers, passengers as well as the industry.

Another important regulation issue to be solved is the spectrum allocation for V2X networking and connectivity technologies. Worldwide regulation administrations have identified a first set of frequency bands for V2X networking and connectivity usage and are expected to further act as the technology evolves. In this context, we will consider three key frequency bands:

- 5.9 GHz: A total of 70 MHz (20 MHz) is allocated to V2X networking and connectivity in Europe and the US and China.
- 3.4 GHz—3.8 GHz: The allocation of 3.4 GHz to 3.8 GHz for IMT usage is under discussion. And the allocation of a part of this band to V2X networking and connectivity is an option.
- 60 GHz: Europe allocated a 1 GHz channel in the 63 GHz to 64 GHz band for V2X networking and connectivity, but the allocation is currently under debate.

In the 5.9 GHz band, the US and European regulation authorities have established an almost harmonized channelization scheme providing 7 vehicular communication channels of 10 MHz each. The allocated band is 5.855 GHz to 5.925 GHz, i.e. 70 MHz in total. In the US, the FCC issued a report and order in 2003 which provides licensing service rules for this band. The intended usage includes in particular V2V and V2I communications, with the objective to protect the safety of the moving public. In China, the allocated resources are still limited, providing a total of 20 MHz bandwidth in 5.905 GHz to 5.925 GHz. For comparison the 3GPP standardization body for wireless cellular communications has defined Band 47 to cover 5.855 GHz to 5.925 GHz as it is illustrated in Figure 3.6.

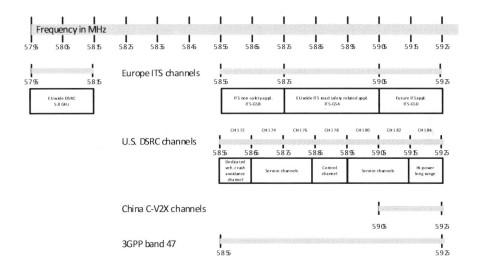

Figure 3.6: Band allocation for V2X networking and connectivity technologies

In the United States, dedicated short range communications (DSRC) is used in the context of IEEE 802.11p based vehicular communications only. In Europe, DSRC is used in the tolling context. Any other vehicular communications services in the spectrum band 5.855 GHz to 5.925 GHz are denoted as ITS-G5. Figure 3.6 shows that the

European regulation anticipates an additional 20 MHz allocation of spectrum for tolling services in the 5.795 GHz to 5.815 GHz range.

To achieve better wireless service reliability and no interruption of safety critical use cases, it is favorable to complement the 5.9 GHz spectrum band with another band that is sufficiently de-correlated. If this spectrum band is sufficiently away from the 5.9 GHz band, it would allow mitigation of such effects as not favorable shadowing effects, poor propagation conditions and interference at least in some or most cases. Therefore the 3.4 GHz to 3.8 GHz spectrum band is another candidate band for IMT usage. There is the anticipation that the band will be allocated for 5G use and is still open to extend V2X networking and connectivity technologies which could make use of this spectrum band. A number of diverging options are available to make use of this spectrum band. From a cellular mobile network operator's point of view, it could be used for V2X networking and connectivity based services bound to cellular network infrastructures, but the usage depends on the outcome of 5G and IMT 2020 standardization. Possible options are likely as follows:

– 3.4 GHz to 3.8 GHz gets fully applied for IMT and network operators could take part of the resources for V2X networking and connectivity.
– 3.4 GHz to 3.8 GHz is fully assigned for IMT but additional constraints are imposed to provide capacity to V2X networking and connectivity.
– Part of the 3.4 GHz to 3.8 GHz band will not be allocated to IMT but rather, kept back for V2X networking and connectivity.

There is the option to use spectrum bands at 60 GHz for V2X networking and connectivity; Europe has got a spectrum band allocation at 63 GHz to 64 GHz which originates from European regulation ECC/DEC/09(01) in 2009 and references outdated ITS system requirements defined in ECC Report 113. Currently, a new spectrum band allocation scheme is under discussion as it will be addressed in the sequel of this chapter.

There are challenges with respect to any of the three band allocations for vehicular communications as discussed above, i.e. 5.9 GHz, 3.4 GHz—3.8 GHz and 60 GHz. To start with the 5.9 GHz band, the corresponding allocation is technology agnostic, i.e. regulation administrations do not mandate the usage of a single specific technology. Currently, two distinct technologies are competing for usage of the spectrum, IEEE 802.11p based DSRC and 3GPP cellular-vehicle-to-X (LTE C-V2X).

These two systems have been defined independently of each other (IEEE 802.11p was defined by IEEE and LTE C-V2X by 3GPP) and no interoperability mechanisms have been defined so far in the respective standards. It would thus be straightforward to apply each of the two standards in dedicated and independent frequency bands, but sharing a single frequency band would be leading to massive interference and system malfunction—the corresponding standards have not been designed for such a usage. For this purpose, the 5G Automotive Association (5GAA) has recently published a position paper, advocating so-called safe harbor channels. In Europe, 5.875 GHz—5.905 GHz (i.e., three channels of 10 MHz each) are allocated to safety related

ITS services. The safe harbor channel approach foresees that the lower and upper band are allocated to IEEE 802.11p DSRC and LTE C-V2X (or vice versa) respectively. See Figure 3.7. The middle channel may be retained for band separation. From a regulation perspective, however, it is being questioned whether such a band split is in alignment with technology neutrality.

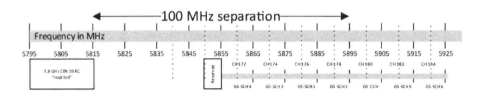

Figure 3.7: Safe harbor proposal by 5GAA

Another possibility consists in the definition of novel coexistence mechanisms which would allow an efficient coexistence of both systems in an identical channel. Then, the requirement of technology neutral operation would be straightforward and might be met.

Another issue relates to the time line and availability of both respective technologies, IEEE 802.11p DSRC and LTE C-V2X. The IEEE 802.11p DSRC standard is stable, related products are available and well tested. The commercial mass-market introduction is planned to start in 2018. LTE C-V2X has not yet reached this level of maturity. 3GPP indeed start with Rel. 14 only to develop vehicular communications features. It will take a number of years until LTE C-V2X is stable; products are available and required field tests have been successfully completed. Without regulation provisions, there is thus the risk that IEEE 802.11p will occupy the entire 5.9 GHz vehicular band and thus make a later introduction of LTE C-V2X difficult. The safe harbor idea of 5GAA, as mentioned above, has the objective of ensuring the possibility to allow for LTE C-V2X to enter the market still at a later point in time, probably in the early 2020s.

Following the reasoning above, LTE C-V2X seems to be under pressure in the short term due to the early stage availability of IEEE 802.11p DSRC products. In the long run, however, the situation is expected to be inversed. As mentioned above, the IEEE 802.11p standard is stable and no further evolution is planned. The 3GPP effort, on the other hand, is planned to transition smoothly towards 5G in a gradual way, ensuring backward compatibility. LTE C-V2X will thus be complemented by new features as the need arises while no evolution will occur for the competing system. Due to this situation, there is wide-spread expectation that 5G vehicular communications will prevail in the long term and finally be the most widely accepted standard. This situation, however, will not occur immediately. The vehicle replacement time is approximately 10 years or more. A full replacement of one standard by another one may only occur after this replacement cycle time.

In the 60 GHz band, the situation is entirely different. In the US, no corresponding allocation exists so far. As mentioned above, in Europe there is a European Regulation decision in place (ECC/DEC/09(01)) which allows the usage of a system designed to operate in the 63 GHz–64 GHz band. No specific vehicular communications system exists so far. This existing allocation in Europe, however, faces challenges. As it is illustrated in Figure 3.8, the band is already allocated to the so-called WiGig (Wireless Gigabit Alliance) system and the 63 GHz–64 GHz overlaps with two such WiGig channels, namely channels 3 and 4.

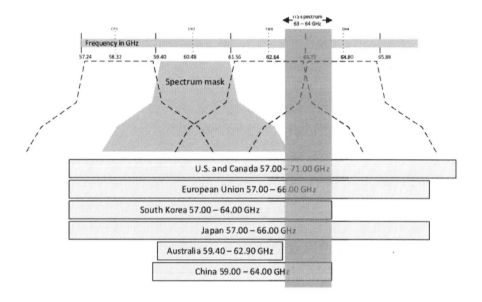

Figure 3.8: ITS spectrum option at 60 GHz

Any usage of Wi-Gig in channels 3 or 4 would thus lead to a prohibitive level of interference onto the ITS system. For this reason, the following proposal is currently being discussed on the level of standards organizations and regulation administrations, either to abandon the 63 GHz–64 GHz allocation or instead re-allocate ITS usage either to Channel 3 (61.56 GHz to 63.72 GHz) or Channel 4 (63.72 GHz to 65.88 GHz).

The proposed re-allocation has a number of advantage that the ITS allocation perfectly matches with a WiGig Channel (either channel 3 or 4). Moreover, the available spectrum is increased from 1 GHz bandwidth (63 GHz to 64 GHz) to 2.16 GHz bandwidth (61.56 GHz to 63.72 GHz for Channel 3 or 63.72 GHz to 65.88 GHz for Channel 4). Thus, more capacity will be available for ITS systems. Furthermore, such a new channelization would allow the usage of existing, off-the-shelf commercial WiGig products for vehicular applications. Corresponding products are likely to be a less expensive solution for the automotive industry compared to a specific 63 GHz to 64 GHz system addressing the vehicular market only. Furthermore, 3GPP is expected to develop alternative solutions which will be designed to fit into the 5G ecosystem. Finally, it is expected that regulation administrations will contemplate the simultaneous usage of WiGig channel 3 and/or 4 for commercial as well as vehicular applications – commercial (non-vehicular) products using these bands are indeed already in the market and further ones will follow. It is possible to introduce a prioritization for vehicular services over other applications. But in the end, such a rule may not be required in practice, since the commercial usage of WiGig for in-vehicle entertainment/communications is expected to be unlikely. Corresponding discussions will certainly dominate related efforts in regulation administration in the future.

Millimeter wave spectrum allocation, including 60 GHz allocation for ITS, will be negotiated at the World Radio Conference 2019 (WRC'19). In case the upper re-allocation is approved by CEPT, it is likely that the new allocation will be Europe's position for WRC'19 discussions.

To summarize, we expect that vehicular communications services will use at least three bands (Figure 3.9): In the 3.4 GHz band, MNOs are expected to provide Vehicle-to-Network (V2N) services, the 5.9 GHz band will be used for Vehicle-to-Vehicle (V2V), Vehicle-to-Person (V2P) and Vehicle-to-Infrastructure (V2I) services while the 60 GHz band finally provides high throughput services. The usage of 60 GHz vehicular communications is still in an early definition stage. The corresponding use cases are still being discussed. Possible applications include high resolution video transmissions in a platooning set-up, high-rate data exchange at localized hot spots and similar.

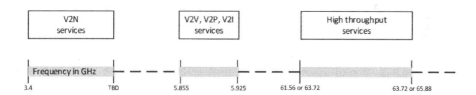

Figure 3.9: Spectrum options for V2X networking and connectivity

3.4 Conclusions

The integration of V2X networking and connectivity with sensor armed vehicles for evolved advanced driver assistance (ADAS) needs to support final solutions for automated and autonomous driving. But at present V2X networking and connectivity technology faces difficulties still to achieve requirements for low latency, reliability, security and functional safety to get implemented for automated and autonomous driving use cases. The vehicle is already one of the devices with the most heterogeneous networking and connectivity when in-vehicle and external radio access technologies like Wi-Fi, NFC, Bluetooth, DSRC and cellular LTE are used. This mix of required multiple radio access technologies contributes to integration challenges as interference mitigation, interoperability and technology life cycles must be resolved.

For sensor extensions in vehicles there are needs to support a huge variety in data bandwidth, from transmitting and receiving several bytes to gigabytes per second including uplink and downlink, intermittent connectivity and a dynamic environment with highly dynamic topology and infrastructure, which makes modeling, simulation and field trial testing a challenge.

Automated and autonomous driving integrates driver, vehicle and environment tightly making environment a key part of the system. Therefore, local dynamic HD maps on board the vehicle have to provide data beyond the line of sight, implementing the eHorizon. Maintaining the maps with real-time data from in-vehicle sensors and sensor extensions due to V2X networking and connectivity means the previously static eHorizon dynamic as well and the vehicle will become able to look beyond several hundred meters and around the corner. Questions in relation to creation, maintenance and use of local dynamic HD maps, like what kind of data and lifecycle, how to ensure sustainability and trust, ownership, service, security, privacy and functional safety still need to be answered going forward.

The V2x networking and connectivity system that provides software and firmware updates over the air (FOTA, SOTA) must provide a high level of security; as high as possible and needs to protect each of the following components of vehicle electronics systems that could be addressed by FOTA and SOTA updates: infotainment, vehicle control, fail safe vehicle body and driving functions and fault functional vehicle and driving functions. Intensive threat analysis must be performed to make the updates tamper proof.

Infotainment is one of the systems in a vehicle, which gains a lot of advantage by integrating evolved V2X networking and connectivity technology. Many infotainment use cases are non-safety but have got a constantly growing demand for high bandwidth. Furthermore, there is the integration necessities of infotainment, navigation, telematics and vehicle control due to at least the use of the human machine interface (HMI). Therefore, requirements of functional safety begin to come into play.

There are differences between V2X networking and connectivity standards and regulation developed in particular in Europe and United States. In Europe the standardization and regulation is driven by the car-to-car communications consortium (C2C-CC), the Amsterdam group, ETSI ITS and the European committee for standardization (CEN/ISO). V2X networking and connectivity standardization and regulation in the United States includes the national highway traffic safety administration (NHTSA), the federal highway administration (FHWA), the department of transportatopm (DOT), the American association of state highway officials (AASHTO) and the

FCC. Harmonization activities happen at the Institute of Electrical and Electronic Engineering (IEEE), the Society of Automotive Engineers (SAE), ETIS, ISO and 5GAA.

The IEEE 802.11p DSRC standard allows reliable low latency communications of basic safety messages between vehicles and between vehicles and roadside infrastructure. Then cellular V2X networking and connectivity standards challenge DSRC standards at a progressive rate. The evolutionary path offered by 3GPP cellular technologies LTE and 5G NR, and their large number of supporting mobile ecosystem stakeholders, make cellular standards an option for vehicle manufacturers and their suppliers. While DSRC is more tested and ready to deploy as of today, it remains to be seen whether it will take off in quantities in the field. Then there are the very demanding requirements of automated and autonomous driving which challenge both standards, DSRC and LTE. At that point, the cellular 5G V2X standardization will surpass DSRC and in the longer term, vehicle ecosystem stakeholders start to look at cellular network-based low-latency, end-to-end V2X networking and connectivity solutions. The main challenge for V2X networking and connectivity is to get hold of sufficient dedicated spectrum by regulation needed for the vehicle to vehicle, vehicle to infrastructure or vehicle to pedestrian communications, tackling the demand for capacity, reliability and functional safety.

References

1609.2-2016–IEEE Standard for Wireless Access in Vehicular Environments – Security Services for Applications and Management Messages, available at https://standards.ieee.org/findstds/standard/1609.2-2016.html

1609.4-2016–IEEE Standard for Wireless Access in Vehicular Environments (WAVE) -- Multi-Channel Operation, available at https://standards.ieee.org/findstds/standard/1609.4-2016.html

3GPP Release 14, available at http://www.3gpp.org/release-14

5G Automotive Association (5GAA), Coexistence of C-V2X and 802.11p at 5.9 GHz POSITION PAPER, 12 June 2017, available at http://5gaa.org/pdfs/5GAA_News_neu.pdf

5GAA-WG4 S-170022, Work Item Description: Study of spectrum needs for safety-related intelligent transportation systems; Huawei, Ericsson, Nokia, Intel, Qualcomm, Vodafone, (February 2017).

802.11p-2010–IEEE Standard for Information technology—Local and metropolitan area networks – Specific requirements – Part 11: Wireless LAN Medium Access Control (MAC) and Physical Layer (PHY) Specifications Amendment 6: Wireless Access in Vehicular Environments, available at https://standards.ieee.org/findstds/standard/802.11p-2010.html

A. Festag, Standards for vehicular communications—From IEEE 802.11p to 5G. Elektrotechnik

AUTOSAR http://www.autosar.org/documents/

Coodination and Support Action, Cooperative ITS Deployment Coordination Support (CODECS), Project Number 653339, Deliverable 2.2: State-of-the-Art of C-ITS Deployment, available at http://www.codecs-project.eu/fileadmin/user_upload/Library/D2_2_CODECS_State-of-the-Art_Analysis_of_C-ITS_Deployment_.pdf

ECC Decision (09)01, Harmonised use of the 63-64 GHz frequency band for Intelligent Transport Systems (ITS), 13 March 2009.

ECC Report 113, Compatibility Studies around 63 GHz between Intelligent Transport Systems (ITS) and Other Systems, Budapest, (September 2007).

ETSI Technical Committee Intelligent Transport Systems, available at http://www.etsi.org/technologies-clusters/technologies/automotive-intelligent-transport

IEEE 1609.3, 1609.3-2016 – IEEE Standard for Wireless Access in Vehicular Environments (WAVE) – Networking Services, available at https://standards.ieee.org/findstds/standard/1609.3-2016.html

IEEE 802.2, IEEE Standard for Information technology – Telecommunications and information exchange between systems – Local and metropolitan area networks – Specific requirements Part 2: Logical Link Control, available at http://standards.ieee.org/about/get/802/802.2.html

IETF RFC 2460, Version 6 of the Internet Protocol (IPv6), sometimes referred to as IP Next Generation or IPng, available at https://www.ietf.org/rfc/rfc2460.txt

IETF RFC 768, User Datagram Protocol (UDP), available at https://www.ietf.org/rfc/rfc768.txt

IETF RFC 793, Transmission Control Protocol (TCP), available at https://tools.ietf.org/html/rfc793

Informationstechnik 132(7), 409–416 (2015). doi:10.1007/s00502-015-0343-0

Intelligent Transportation Systems Using IEEE 802.11p; Application Note; Rohde & Schwarz, 2014; available at http://www.rohde-schwarz-usa.com/rs/rohdeschwarz/images/1MA152_ITS_using_802_11p.pdf

Intelligent Transportation Systems Using IEEE 802.11p; Application Note; Rohde & Schwarz, 2014; available at http://www.rohde-schwarz-usa.com/rs/rohdeschwarz/images/1MA152_ITS_using_802_11p.pdf

International Organization for Standardization, Road Vehicles—Functional Safety. ISO 26262, 2011 (2011).

Leading the world to 5G: Cellular Vehicle-to-Everything (C-V2X) technologies, Qualcomm, June 2016, available at https://www.qualcomm.com/media/documents/files/cellular-vehicle-to-everything-c-v2x-technologies.pdf

Leading the world to 5G: Cellular Vehicle-to-Everything (C-V2X) technologies, Qualcomm, June 2016, available at https://www.qualcomm.com/media/documents/files/cellular-vehicle-to-everything-c-v2x-technologies.pdf

RP-161919, Liaison Statement from 3GPP RAN on LTE-based vehicle-to-vehicle communications. This document informs internal 3GPP and external stakeholders that the specification work on V2V support using LTE sidelink is completed. Available at http://www.3gpp.org/news-events/3gpp-news/1798-v2x_r14

SAE International On-Road Automated Vehicle Standards Committee, Standard J3016: Taxonomy and Definitions for Terms Related to On-Road Motor Vehicle Automated Driving Systems (2014).

SAE J2735, Dedicated Short-Range Communications (DSRC) Message Set Dictionary™, available at http://standards.sae.org/j2735_200911/

SAE J2945, Dedicated Short-Range Communications (DSRC) Common Performance Requirements™, available at http://standards.sae.org/wip/j2945/

Status of the Dedicated Short-Range Communications Technology and Applications Report to Congress www.its.dot.gov/index.htm Final Report—July 2015 FHWA-JPO-15-218

U.S. Federal Communications Commission (FCC), Dedicated Short-Range Communications (DSRC) Service, available at https://www.fcc.gov/wireless/bureau-divisions/mobility-division/dedicated-short-range-communications-dsrc-service

Wireless Gigabit Alliance, see https://en.wikipedia.org/wiki/Wireless_Gigabit_Alliance

Chapter 4
Technologies

Today in more and more vehicles the sharing of computing, digital storage and net-working and connectivity technology grows steadily. What makes vehicles so attractive for this growth in particular are the chances these technologies provide for example from in-vehicle infotainment, navigation and telematics, vehicle and engine control to parking and driving assistance. More and more innovations in the vehicle account of hardware and software-controlled computing platforms and networking and connectivity. Electric mobility, autonomous and highly automated fully net-worked driving vehicles are currently the prevailing topics of many developments. These vehicles are equipped with high-performance data processing platforms for sensors, computing, storage, actuators and networking and connectivity (Figure 4.1).

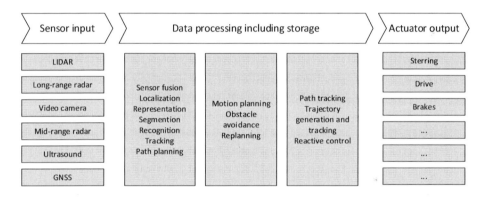

Figure 4.1: Data processing in autonomous and automated vehicles

Autonomous and automated vehicles count on a great deal of sensors, from ultra-sound, video cameras, radar up to LIDAR. These sensors provide massive amounts of fused data about the environment around the vehicle. Since for example a video camera delivers valuable data at good daylight, radar or LIDAR senses at night or provides depth data. Depth data are indispensable to enable for instance the recognition of object differences between pedestrians and trees or traffic signs. In conclusion there are terabytes of data per vehicle which needs to be crunched through or up- and downloaded every day. The entire V2X system must collaborate cooperatively and interconnected, and each node requires a capability to process, store and connect significant amount of data. The emphasis is on data download as well as upload, since gathered data enables the vehicles, infrastructure components and the cloud to adapt and learn from previous experiences and environments.

DOI 10.1515/9781501507243-004

Various sensor technologies such as radar, LIDAR, ultrasonic sound and video camera systems analyse the surrounding vehicle environment. These sensors produce several terabytes of data, which need to be stored and processed by high-performance computing platforms. The vehicle must be linked internally with other units and externally with other vehicles and with the infrastructure of various vehicle ecosystem stakeholders residing at data centers in the cloud. Many computing and communications technologies are instrumental in the digitalization of the vehicle ecosystem in particular for autonomous and automated driving.

The computing hardware architectures for SAE level 4 and 5 vehicles perform the pre-processing and the fusion of the gathered sensor data, the learning and decision making, the HD maps updates, the networking and connectivity and the vehicle actuation control. Several video cameras, radars, ultrasonic sound sensors and LIDAR's deliver Gigabits per second and sensor fusion becomes the norm requiring fully programmable flexible and scalable power efficient computing platforms together with smart data storage systems for data logging and recording, maps and infotainment. Artificial intelligence and machine learning become a vehicle option running algorithms for object detection, path identification and planning requiring high performance computing platforms as well. The vehicle implementing V2X networking and connectivity becomes an enabler for mobility as a service with storage, communications and networking capabilities, data and signal processing ability and sensing. The vehicle perception and localization rely on the aggregation of various compute and communications intensive functions into one powerful computing platform.

The automated and autonomous driving vehicle software runs on top of hardware platforms with defined operation domains like perception, localization, vehicle behavior and control, networking and connectivity, infotainment and equivalent controllers. Software in vehicles expands dramatically in the ballpark of several hundreds of millions of lines of code. The increasing system complexity of automated and autonomous driving vehicles requires an increasing need for frequent, seamless and quick software updates impacting all vehicle ecosystem stakeholders. The trend from hardware to software defined systems challenges the way software gets updated and needs to be adapted to vehicle ecosystem needs. Firmware over-the-air and software on-the-air (FOTA, SOTA) become a necessity. Additional contests are software safety and security, the highly dynamic environment and efficient cloud based vehicle software management.

A vehicle of any SAE level are comprised of computing units for infotainment, with telematics and navigation, vehicle control and connectivity. Currently we see the development of today's vehicle navigation, telematics and infotainment from standard navigation maps with TMC traffic data, GNSS positioning and annual map updates together with ADAS systems moving toward automated and autonomous driving with high reliable ADAS, HD maps including live data overlays and very accurate positioning and localization. These self-learning HD live maps get updated in real time and are cartographed by all automated and autonomous vehicles equipped with corresponding vehicle connectivity to the data centers in the cloud.

Networking and connectivity platforms get integrated with vehicle control platforms and many vehicle functions get connectivity for infotainment, telematics, navigation and vehicle control. Vehicle ecosystem cyber security and privacy are the major prerequisite to prevent attacks on functional safety and preclude hacks on ADAS features such as adaptive cruise control (ACC), pre-crash systems and automatic parking. Functional safety is not only about the vehicles; it includes traffic infrastructure, navigation and telematics and infotainment as well.

4.1 Sensing

The average number of sensors in today's vehicles is over 100 and growing significantly. Sensors and actuators are implemented for active and passive safety, convenience, infotainment, low emissions, energy efficiency and cost and weight reduction. Sensors and actors are networked and connected with engine control units (ECU) using several different in-vehicle networks. Vehicular sensors measure position, pressure, torque, exhaust temperature, angular rate, engine oil quality, flexible fuel composition, long-range distance, short-range distance, ambient gas concentrations, linear acceleration, exhaust oxygen, comfort/convenience factors, night vision, speed/timing, mass air flow, and occupant safety/security. Sensors and actors are in the powertrain and chassis control (engine, automatic transmission, hybrid control, steering, braking, suspension), body electronics (instrument panel, key, doors, windows, lighting, air bag, seat belts), infotainment (audio, video, speech, navigation, traffic message channel (TMC), electronic toll collection (ETC)) and other assistance systems like electronic stability control, pre-crash safety, park or lane assist.

There are currently in-vehicle sensors for temperature, pressure, power, flow, fill level, distance or angle. Temperature sensors are used for oil, cooling water, exhaust gas and the inside or outside of the vehicle. Pressure sensors are for oil, cooling water, tire pressure or hydraulic oil. Power sensors are implemented for airbags, belt tensioners and pre-tensioners, window lifters, brakes and others. Flow sensors are used for oil. Fill level sensors are deployed for gas, cooling water or hydraulic oil. Angle sensors are applied for ABS, ESP, steering, pedals, damper compression or chassis inclination. More specific sensors are knock sensors, acceleration and vibration, engine speed, velocity, lambda probe, linear plate, brightness, humidity, torque, magnetic field and air mass.

All vehicle sensors have rigorous real-time and safety requirements. For example, an anti-lock braking system (ABS) measures the vehicle speed and rotational speed of the vehicle's wheels to detect skid. When skid is detected the pressure to the brake is released to stop the skid, but a permanent reduction in fault case has to be circumvented. Another instance is an airbag control (AC) system which monitors various vehicle sensors including accelerometers to detect a collision. If a collision is detected, the ignition of a gas generator propellant will be triggered to inflate a bag. The trigger for the ignition must be within 10 to 20 milliseconds after the collision. Here the functional safety requirements are even tougher than ABS.

To get to autonomous or automated vehicles, which shall perform human driving capabilities, it all starts with the perception of the vehicle surrounding environment by gathering huge amounts of data in real time by different sensors. It is essential for these vehicles to know their exact position, to estimate perfectly where to go safely next and to control extremely well how to get there. But there is more when it comes to vehicle environment awareness as kind of another essential data set. These additional data are coming from the cloud infrastructure and other vehicles. All this vehicle surround sensing must be highly robust and reliable in all use cases. Since sensors have got their specifics and work differently (see Table 4.1), there is no one sensor technology which fits all use cases.

Table 4.1: Set up for sensing

Scenario	Ultrasonic sound	Radar	LIDAR	Video camera
Rain	Works	No problem	Issues	Issues
Snow	Works	No problem	Issues	Issues
Night	Works	No problem	Works	Issues
Heavy sunlight	Works	No problem	Issues	Issues
Object recognition	Issues	Issues	Works	Works
Distance	Up to 5 meters	Up to 200 meters	Up to 150 meters	Up to 100 meters

Vehicle environment sensor implementations currently include ultrasonic sound sensors, near, mid- and long-range video cameras, mid- and long-range radar or LIDAR. A sensor package for a SAE level 3 vehicle includes up to 12 ultrasonic sound sensors, up to 5 video cameras, up to 3 radar sensors and at least one LIDAR sensor (Figure 4.2). In the case of the Tesla it works with 4 to 8 surround video cameras to provide 360 degrees of visibility around the vehicle at up to 250 meters of range. 12 ultrasonic sound sensors complement the vision sensors, allowing for detection of both hard and soft objects at up to 2 meters distance. A forward-facing radar provides additional data about the environment on a redundant wavelength that is able to see through heavy rain, fog, dust and the vehicle ahead.

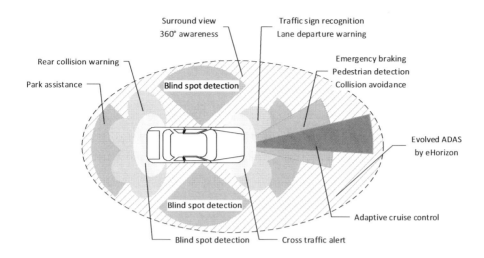

Figure 4.2: Vehicle sensors supporting ADAS use cases

An ultrasonic sound sensor sends out sound waves. When the sound waves hit an object, they produce echoes revealing the obstacle location. The sensor works for soft and hard objects being in a horizontal angle of up to 120 degrees for distances up to 5 meters. Vehicles implement ultrasonic sound sensors to detect obstacles like pedestrians, vehicles or posts in the immediate vicinity of the vehicle. For larger distances and more details, multiple video camera sensors deliver images of the vehicle's surroundings. Rear view and near range video cameras deliver a detection range up to 15 meters with a horizontal aperture of around 130 degrees. A night vision video camera sensor looks up to 150 meters ahead with a horizontal aperture angle of around 30 degrees. And the front video and stereo-video cameras have a detection range of up to 80 meters with a horizontal aperture angle of around 40 degrees.

Video camera sensors detect colors, signs, characters, textures and are therefore capable of traffic sign, traffic light or lane marking detection. Stereo cameras and 3D cameras with depth image provide 3D vision making range determination possible. Furthermore, video camera sensors are used as well for central mirror and external side mirrors or to optically sense drivers in-vehicle. But cameras have got issues with increased distance up to 250 meters to enable anticipatory driving or providing reliable data on weather conditions such as fog, rain, snow, LED flicker of surrounding vehicles' headlights or sun light glare effects.

Therefore, radio detection and ranging (radar) sensors with increased range, angular and elevation resolution, complementing video camera sensors, get added. Radar sensors provide good ranging ability and relative velocity measurement and operate under any weather conditions and in harsh environments including dust, dirt, light, rain, snow, and so on. Radar sensors transmit electromagnetic waves. If radar waves get reflected from objects, they will reveal the objects' distance and velocity. Mid- and long-range radars are deployed around the vehicle to track the distance and speed of objects like vehicles, motorcycles, bikes and pedestrians in the neighborhood of the vehicle in real-time. Radars scan horizontally delivering 2D data. If a vertical scanning is added, you will get 3D data. The vehicle front and rear mid-range radar provides a detection range of up to 150 meters with a horizontal aperture angle of around 45 or 130 degrees. The vehicle long range radar range is up to 250 meters with a horizontal aperture angle of 30 degrees. To increase the radar sensor 3D performance even further, to track detected objects and classify them, the elevation, range, Doppler and angular resolution needs to be optimized for radar frequencies up to 79 GHz.

Tackling the high-resolution required for automated and autonomous driving, light detection and ranging (LIDAR) sensors scan the environment with a non-visible laser beam, which measures distances and produces a full 3D image of the vehicle's surrounding. LIDAR sensors combines one, or more, lasers with a detector that senses the photons reflected from scanned objects, along with built-in data processors that measure the time of flight (ToF) to detect structure and motion in three dimensions. Whereas a single fixed laser performs simple ranging, advanced LIDAR systems use multiple lasers or a rotating system that scan much further and provide wider fields of view.

Together with sensor data fusion, LIDAR allows a high resolution vehicle location estimate, including the positions of surrounding vehicles, pedestrians and other objects. And flash LIDAR, where a single laser pulse illuminates all pixels per frame and the laser pulse return is focused through the lens onto a 3D focal plane array to do imaging through obscuration, provides an even significantly higher resolution vision around the entire vehicle.

Out of these sensor data a comprehensive real-time environment model including traffic participants (cars, pedestrians),the static environment (occupancy grid), a road model (geometry, conditions), traffic control data (speed limits, traffic lights) and precise map localization (landmarks), based on sensor data fusion of continuous, sensing ultrasonic sound, video cameras, radars and 3D flash LIDAR, along with the vehicles precise location, enables autonomous and automated driving of level 3 and above. In-vehicle road condition sensors like video camera, vehicle dynamics control (anti-lock braking system, traction control system, active yaw control), tire effects (slip, vibration), local vehicle weather (air temperature, rain intensity) and cloud data (digital weather maps, dynamic safety maps) get added and all data gets fused into a fusion model including environment, vehicles (dynamics model) and tires (friction estimation).This temporal model is then applied for example to ADAS use cases as shown in Table 4.2 to sense the road, curb and lanes to determine the center of the lane, lane edge, road curb and correct GNSS errors.

Table 4.2: Vehicle sensor data fusion for ADAS

Sensors	Adaptive cruise control /ACC)	Forward collision warning system (FCW)	Auto emergency braking (AEB)	Traffic sign recognition (TSR)	Lane departure warning (LDW)	Rear cross traffic alert (RCTA)	Active park assist (APA)
Radar 24 or 77 GHz	Yes	Yes	Yes			Yes	
Video mono or stereo	Yes	Yes	Yes	Yes	Yes		Yes
Video rear view					Yes	Yes	
LIDAR short		Yes	Yes				
LIDAR far	Yes	Yes	Yes				
Ultrasonic sound							Yes
Radar and video mono	Yes	Yes	Yes	Yes	Yes		
Radar and video stereo	Yes	Yes	Yes	Yes	Yes		
Radar and video mono and ultrasonic sound	Yes	Yes	Yes	Yes	Yes		
Radar and HD map	Yes						
Video and HD map	Yes			Yes			
Radar and video and HD map	Yes	Yes	Yes	Yes	Yes		

The requirements on data throughput, latency and reliability of V2X networking and connectivity for sensor technology depend very strongly on the traffic use cases and in particular their very diverging dynamics. Because of the vehicle and traffic scenarios dynamics, the communications requirements depend among other parameters on the communication distance, vehicle speed and number and density of devices, which needs to get linked with each other. For example, use cases with the need for cooperative perception or cooperative decision making for lane or parking assist require a much higher number of traffic participants to get linked than use cases like adaptive cruise control or traffic sign recognition.

There are use cases where the exchange of raw, pre- or post-processed vehicle sensor data is needed, although it consumes much more networking and connectivity resources. For example, there are algorithms where the processing of data for the same objective is not the same on different vehicles. Different algorithms process the same raw data for a different purpose, e.g. lane keeping assist versus vehicle following. Advantages of the exchange of pre- or post-processed data instead of raw sensor data include lower bandwidth consumption and scalability.

Another strong demand for V2X networking and connectivity comes from the evolution of vehicle sensing to comprehensive 3D perception as required for SAE level 3 and above use cases. Insufficient communications and computing performance, high cost and poor precision have prevented 3D systems in many vehicle use cases. Now 3D perception technology is gaining momentum in more and more use cases through increased performance and high-resolution sensors. There are various technologies to gather three-dimensional distance data from a traffic scenario. Active solutions such as LIDAR or other time-of-flight sensors determine distance information. These solutions require a decent amount of computing performance and have almost no constraints on the traffic scenario structure. The resolution of current time-of-flight systems depends on the kind of sensor used and their use in real traffic scenarios relies on good viewing conditions. Passive solutions use image data taken by video cameras, similar to the distance perception of the human visual perception system. They offer good spatial resolution, but require high performance computing and suffer from not so good lighting conditions and high textured traffic environments. Data pattern projections and video camera stereo systems enable high spatial and depth resolutions.

4.2 Computing

Depending on the type of vehicle, ranging from small, compact, intermediate, large sports utility and luxury vehicles up to vans, the number of hardware components varies between 10 and 150. Vehicles today contain around 100 electronic control units (ECU) or more when all optional features are chosen when buying a vehicle. An ECU is a computing platform comprised of one or more micro controller units (MCU) to control a certain system domain and vehicle function by using input sensors and actuators. There are ECUs for engine systems, transmission control, electric power steering (EPS), hybrid electric vehicle (HEV), break-by-wire systems, airbags, smart keyless, tire pressure monitoring system (TPMS), dashboard, adaptive front lighting systems (AFS), body control, doors or wipers. For instance, the engine system ECU processes data from vehicle sensors like crank position, an air flow meter, intake temperature and throttle sensors to control fuel injection volume, ignition timing and many vehicle actuators. The real-time behavior is determined by rotation speed and motor cycles, for example at 6000 rpm one cycle is 20 milliseconds. The real-time requirement for ignition is in 10 µs order, so the processing of sensor data for the fuel injection volume has to be finished within 10 µs.

Vehicle networking and connectivity between ECUs enable the sharing of data and avoid redundancy of sensors, actuators, storage and computing building blocks. The in-vehicle system architecture evolves from a central gateway architecture with up to 100 ECUs in one vehicle and one ECU per function and signal based communications to a domain controller architecture with dedicated and secure functionality domains and service oriented networking and connectivity. In the future we might see a centralized functional architecture with virtualized data processing and a networking and connectivity supporting computing domains (Figure 4.3).

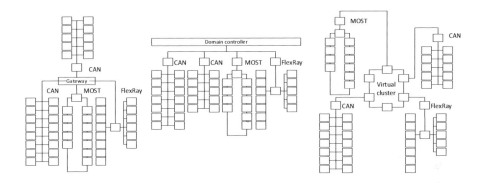

Figure 4.3: From central gateway (left) to domain controller (middle) and functional architecture (right)

The recent computing platform advancement challenges vehicle manufacturers and the whole ecosystem. While vehicle product cycles spanning at least three years or more, vehicle manufacturers and suppliers struggle to keep up with the much faster evolution of computing platforms; for example, being used in smartphones whose life cycles are very often shorter than one year. And several computing platforms get already integrated into vehicles across IVI, driver assistance, automated and autonomous driving, and other functions and features. Doing so, vehicles become more and more software-defined, exploiting the computing platforms advancement, while dealing with the challenges of safety, security, and system costs at the same time.

In order to cover these challenges in the vehicle, functions get consolidated into a number of domain or area controllers that evolve from today's complex architecture that is based upon a huge number of electronic control units (ECUs) distributed throughout the vehicle. Therefore, the vehicle ecosystem stakeholders look for a flexible and scalable vehicle computing platform architecture that can be configured to support various different functions by varying the number and configuration of domain or area controllers in the vehicle and the software that run on them.

The next generation of advanced driver assistance systems (ADAS) technology uses sensors which generate Gigabytes of data that need to be fused, analyzed and processed fast enough so that decisions made and vehicle actions are performed in real-time. The vehicle has to get environment awareness to adapt to never seen before evolving scenarios in less than a second based on huge amount of input data such as vehicle neighborhood objects, speed, road, or weather conditions. The vehicle must account for the unpredictable behavior of other vehicles, pedestrians and other objects while steadily communicating.

There are different approaches for computing platforms targeting ADAS and autonomous and automated driving use cases. For example, ADAS use cases like adaptive cruise control (ACC) or lane keep assist (LKA) are implemented as a distributed data processing architecture where dedicated ECUs process sensor data specific for one ADAS function creating latency and connectivity challenges for the central computing unit for sensor data fusion. Another more scalable and flexible option is a central unit which is capable of raw sensor data processing of every sensor as well as of doing sensor data fusion without latency to support ADAS use cases.

One approach for automated and autonomous driving depends on pure computing performance to deal with the overwhelming amount of sensor data. In this approach for instance video cameras, radar and LIDAR data are fused with high-definition maps (HDM), data to estimate the vehicle location. Another approach gathers only as much sensor data as necessary to keep the amount of data manageable. It gathers video camera and radar data to recognize outlines and textures of the environment. Then it applies for example triangulation to determine the position of the vehicle between two precisely measured waypoints.

There are options to build the computing platform on a GPU, communications or server centric platform in implementing SAE level 3 or above for automated and autonomous-driving. The computing platform needs the processing performance to process data from quite a lot of video cameras, radar and ultrasonic sound sensors as well as the LIDAR sensor and is dedicated to supporting neural-network inferencing as a deep-learning accelerator delivering trillions of operations per second. The data processing on the platform supports all the required functions for use cases of SAE level 3 and in the future above.

And another basic data processing approach is rapidly changing. Plenty of data get uploaded to data centers in the cloud, which apply deep artificial intelligence, deep and machine learning algorithms to them. The resulting instructions and rules are then transferred back to the vehicles, telling them what's what in their neighborhood and what the vehicles should deal with. Vehicles start to improve their recognition rates of surrounding objects, for example everything from upcoming stoplight, chatting pedestrians, or a recent collision ball running into the road. There are many data-hungry processing, storage and communications steps along with the many vehicles networking and connectivity links.

Autonomous and automated driving is more than simply gathering and processing huge amounts of sensor data and fusing them in an environment model and feed them into vehicle system domain ECUs. Every vehicle has a need of a superior understanding and full awareness of every feasible traffic situation, which enables the vehicle control system to decide in any quickly changing circumstances. Deep and machine learning and artificial intelligence technology might have the potential to deliver and call for high performance server computing for vehicles, the cloud and a high bandwidth communications interfaces between both.

Deep neural networks (DNN) process sensor data through successive layers, while each layer applies multiple linear and non-linear operations. DNN training uses offline huge demonstrative data sets, whereas the applied trained system then processes the sensor data in real-time. The training vehicles implement a shadow mode, where the vehicle runs the computing platform with actual sensor inputs and records the system outputs together with driver outputs to gather and transfer huge amounts of data. All use cases get captured and the DNN trained offline and validated afterward in the cloud. When the new functions are proven they get enabled via FOTA and OTA software update.

For comparison, convolutional neural networks (CNN) implement several convolutional layers and a lower number of fully-connected layers. The number of convolutional layers depends on use cases and is different with up to 250 layers. CNN uses a special architecture containing feature maps or convolutions where this architecture is adapted to recognize in particular images. Recurrent neural networks (RNN) are used to process sensor data which are changing over time for example for motion planning. A RNN samples therefore data collections over time.

4.3 Communications

There are two implementation areas where V2X communications has got an impact, intra-vehicle networking and connectivity and outside vehicle networking and connectivity. There is a steadily rising request for data connectivity inside vehicles. The requirements of in-vehicle networking and connectivity are very diverse according to their application area and use cases. Thus, vehicle manufacturers apply networking technologies like the controller area network (CAN), the local interconnect network (LIN) and FlexRay to connect electronic control units (ECUs) with each other.

CAN is currently the de-facto standard protocol for in-vehicle networking with several Mbps bandwidth and non-deterministic behavior under high load. It is widely used in the basic trunk network, in the powertrain and body systems. The current flexible data-rate protocol (CAN FD) delivers communications up to 5Mbit/s and with partial networking (CAN PN) at an improved energy efficiency. LIN is a single-master in-vehicle networking bus protocol vehicle for body applications with 19.2 kbps and a UART interface that is used in switch input and sensor input actuator control. FlexRay is a high-speed, high-performance communications protocol up to 10 Mbps, with flexibility, reliability and security which is mainly used for X-by-wire, ADAS, and high performance applications. And the media oriented systems transport protocol (MOST) is designed for multimedia networking using optical fibre with bandwidth up to 150 Mbps.

For example, the chassis network applies CAN or FlexRay for high data rate, guaranteed response time and high reliability whereas the body electronics network implements CAN or LIN to link a large number of ECUs at moderate data rates and reliability but with low power consumption. The infotainment network applies MOST for multimedia high data throughput. In-vehicle networking and connectivity next-generation technologies include options such as mobile high-definition link (MHL), high-definition multimedia and control over a single cable (HDBaseT), Wi-Fi, near field communications (NFC) and Ethernet which is mainly used for diagnostics today but has high potential for more.

The intra-vehicle and outside vehicle communications technology evolves in stages. In the first stage ECUs get applied independently to various vehicle components like the engine, brakes, steering, etc. and the in-vehicle network is not used. In the next stage each ECU exchanges data for improving the efficiency of the control system and each system operates almost independently. In this stage the timing constraints on vehicle networking and connectivity are loose. With integrated systems in the next stage, each system still operates autonomously, but some applications are provided with multiple ECUs connected with in-vehicle networks. For functional safety the mechanical backup system is still present and the basic functions of a vehicle are well-maintained even if the electronics system fails. In the next stage V2X networking and connectivity are intensively implemented and the mechanical systems get exchanged with ECUs and networks.

The wireless V2X networking and connectivity technology has to fulfill the often contradictory requirements of all vehicular use cases with their extreme complexity of network topologies, mobility, environment dynamics and technology heterogeneity. It has to support unidirectional broadcast and geocast as well as direct communications peer-to-peer in ad-hoc or direct mode with always available coverage.

Wireless V2X networking and connectivity technologies are expected to provide traffic and transport improvements for safety and traffic efficiency applications. Using ad-hoc wireless communications, a variety of data are exchanged between vehicles or with the vehicle infrastructure. The major wireless technologies for V2X networking and connectivity are the wireless access for vehicular environments (WAVE), SAE J2735, ITS-G5 and 3GPP at present. Wireless access for vehicular environments (WAVE) is an approved amendment to the IEEE 802.11 standard also known as IEEE 802.11p. WAVE makes sure the traffic data collection and transmission are immediate and stable, and keeps the data security applying IEEE 1609.2. IEEE 1609.3 and defines the WAVE connection setup and management. The communication between vehicles (V2V) or between the vehicles and the roadside infrastructure (V2I) is specified for spectrum in the band of 5.9 GHz between 5.85–5.925 GHz.

The spectrum allocation for WAVE started with the request to allocate 57 MHz of spectrum in this 5.9 GHz band for intelligent transportation systems (ITS) in the United States in 1997. In October 1999, the FCC allocated the 5.9 GHz band for DSRC-based ITS applications and adopted basic technical rules for DSRC operations. The

Federal Highway Administration (FHWA), an agency of the USDOT, developed a national, interoperable standard for dedicated short range communications (DSRC) equipment operating in the 5.9 GHz band until 2001. IEEE 802.11p has been tested in many projects and large number of vehicle ecosystem stakeholders at may sites worldwide.

Another wireless technology that is commonly known in vehicular communications and, in particular, V2V networking and connectivity is the J2735 DSRC message set dictionary, maintained by the society of automotive engineers (SAE). This standard specifies a message set, the data frames, and data elements specifically for applications anticipated to use DSRC/WAVE communications systems. The message set is comprised of 15 messages, 72 data frames, 146 data elements and 11 external data entries. Message types are basic safety, a la carte, emergency vehicle alerts, generic transfer, probe vehicle data and common safety requests. In Europe, ETSI defined ITS-G5, which incorporates some of specifications from WAVE.

Wireless communications technologies complement on-board vehicle sensors (Figure 4.4) and increase the vehicles' perception beyond line of sight (eHorizon). Cooperative awareness messages (CAMs) are continuously sent out by all vehicles in a specific region. These messages contain the transmitter's location, speed, and direction of travel and allow for use cases like intersection collision avoidance (ICA) to warn drivers if other vehicles are detected that are on a collision course. Signal phase and timing (SPaT) messages sent out by traffic lights enable the green light optimal speed advisory (GLOSA) use case, where drivers can adapt their speed. Event-based decentralized environmental notification messages (DENMs) are forwarded over several hops and are used to warn vehicles of hazardous situations such as the end of a traffic jam on the highway.

Improvements of 3GPP technologies of the already standardized LTE-Direct standard (ProSe, 3GPP release 12) lead to the V2X specifications in 3GPP release 14 and address safety critical and non-safety applications in common activities with standardization bodies ETSI ITS WG1 and SAE DSRC. There is an objective that the LTE V2V solution should be able to use V2X messaging protocols like SAE J2735 and cooperative awareness messages (CAM), decentralized environmental notification messages (DENM), signal phase and timing (SPAT), map data (MAP) and IVI.

Depending on the considered use-cases, distinct requirements come into play. Applications for in-vehicle infotainment (IVI) require high bandwidth and network capacity, active road safety relies on delay- and outage-critical data transmission, whereas data exchange for road traffic efficiency management typically comes without strict quality of service (QoS) requirements and exhibits graceful degradation of performance with increasing latency.

3GPP considers link types for V2X networking and connectivity use cases such as vehicle to vehicle (V2V) which it treats with an assisted inter-vehicle data exchange, vehicle to network (V2N) which delivers connectivity to the cloud, vehicle to infrastructure (V2I) which covers wireless links to RSUs and vehicle to pedestrian (V2P) which supports data exchange with other devices in the proximity of vehicles e.g. pedestrians and cyclists. Dual connectivity in dense heterogeneous networks, LTE-based broadcast services and enhanced multimedia broadcast and multicast service (eMBMS) and proximity services including D2D communications are used to implement LTE based V2X networking and connectivity.

V2X networking and connectivity is capable of enabling many use cases. But unless the functional safety challenge is not solved, all automated and autonomous driving use cases are out of scope, and the use of V2X is mainly for mobility as a service, convenience or infotainment related use cases. These use cases do not rely on embedded connectivity and therefore are possible with smartphone mirroring and tethered connectivity to get vehicle ecosystem stakeholders linked.

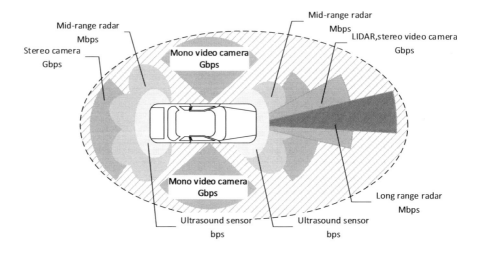

Figure 4.4: V2X networking and connectivity data rates by sensors

Latency rises with higher numbers of vehicles in a network cell. For instance, it has been shown (Phan, Rembarz, & Sories, A Capacity Analysis for the Transmission of Event and Cooperative Awareness Messages in LTE Networks, October 2011) that with LTE technology, the capacity limit for distributing event-triggered messages to all devices in the same cell can reach up to 150 devices in urban scenarios and about 100 devices in rural scenarios, maintaining an end-to-end delay below 200 milliseconds.

Every data packet (e.g., between two nearby vehicles) must traverse the infrastructure, involving one uplink (UL) and one downlink (DL) transmission, which may be suboptimal compared to a single radio transmission along the direct path between source and destination nodes, possibly enjoying much lower delay (especially in an overloaded cell). In addition to being a potential traffic bottleneck, the infrastructure may become a single point of failure e.g., in case of eNB failure.

3GPP LTE is originally designed for broadband traffic and is not optimal for transmitting small amounts of data for any V2X use case. This turns into suboptimal usage of radio resources and spectrum. However, many V2X use cases require support for a large number of very small-sized packets. This leads to potential issues within the current cellular designs, which are, for example, channel coding, radio resource management or control and channel estimation. In particular, commonly implemented control and channel estimation quickly becomes very inefficient for very short payloads.

Another issue for many vehicle use cases is the protocol delay until a payload can be transmitted. The common random access procedure in LTE is a multi-stage protocol with several messages in both uplink and downlink. Even a simplified implementation of the existing LTE access requires at least one preamble transmission and one downlink feedback preceding the payload transmission due to required uplink synchronization. The set of at most 64 preambles per sub-frame are shared among all devices in a cell, regardless of their applications. The preambles are utilized for initial access, re-synchronization for data transmissions, handover, and radio link failure recovery. Therefore, the preamble scheme has many constraints with respect to delay and Doppler spreads, which limit the spectral efficiency of the physical random access channel (PRACH) and does not scale with the number of devices. In addition, an increasing number of random access responses, which are 56 bits per device, restricts the overall downlink capacity further.

Another drawback of the wireless cellular LTE infrastructure-based approach is that it is not available out of coverage and does not satisfy the stringent functional safety needs of vehicles therefore. This implies that additional infrastructure or specific standard extensions must get deployed if coverage is to be guaranteed.

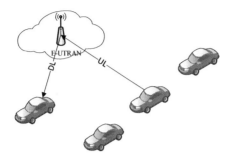

Figure 4.5: V2X operation only based on Uu interface

Figure 4.6: V2X operation only based on PC5 interface

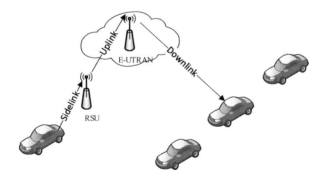

Figure 4.7: V2X operation using both PC5 and Uu interface

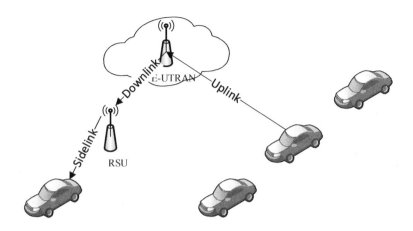

Figure 4.8: V2X operation using both PC5 and Uu interface

Table 4.3: Key performance indicators for V2X networking and connectivity

Feature	IEEE 802.11p	LTE	5G NR
Type	Dedicated	General purpose	
Market penetration	Low	Low	high
Max bit rate in Mbps	27	300	
End-to-end delay in milli-seconds	10	10	1, 2, 5 less than 10
Set up time in seconds	0	100	
Max range in km	1000	30	
Coverage	intermittent	ubiquitous	ubiquitous
Mobility support	medium	high	high
V2V broadcast	Yes	Server	
V2V multi hop	Yes	Server	
V2I bidirectional	Yes	Yes	
I2V broadcast	yes	eMBMS	

4.4 Software

An automated and autonomous vehicle system is a very complex hardware and software system. The amount and complexity of software in vehicles increases, and the number of lines of code grows further, into the hundreds of millions of lines of code ballpark, controlling hundreds of electronic control units (ECU). More functions, more interfaces, more V2X networking and connectivity, more sensors and actors, more human-machine-interfaces and more diversity in the supply chain create huge challenges. And the very different software development and innovation cycles of the vehicle and the computing and communications stakeholders must be aligned. Additionally, the software has to comply with functional safety and must be maintained very efficiently.

There is a rush goin on, lead by major computing and communications stakeholders in collaboration with vehicle manufacturers, to develop and to mature the autonomous and automated driving hardware and software stack. SAE level 4 software development is under way for several platforms and vehicles, test driving in urban and rural environments. Part of the work is in support of computing platforms with field programmable gate arrays (FPGA), GPU, memory, sensors, recurrent neural networks (RNN) and convolutional neural networks (CNN) providing localization and planning with data path processing, decision and behavior with motion, behavior modules and arbitration, control with the lockstep processor, safety monitors, fail safe fallback and X-by-wire controllers as well as V2X networking and connectivity.

The automotive open system architecture (AUTOSAR) is a consortium composed mainly of vehicle manufacturers and electrical equipment suppliers with the objective of developing a common-use automotive software and electronic architecture by standardization of software platforms and development processes, relating to in-vehicle networking such as CAN, LIN and FlexRay. The current AUTOSAR software architecture for electronic control units which is primarily developed for vehicle control functions gets developed further to support more real-time data processing and networking and connectivity required for autonomous and automated driving.

AUTOSAR work driven by vehicle manufacturers puts emphasis on the software architecture, the software development methodology and the application interfaces to support software for different functional domains containing ADAS, vehicle control and FlexRay, CAN or Ethernet. The runtime environment (RTE) servers as software abstraction layer between hardware independent application software from vehicle computing platforms and architecture dependent software (Figure 4.9). The corresponding software modules are independent and are used flexibly in relationship with the application interfaces which are available for the vehicle's body, interior and comfort, power train, chassis and passenger and pedestrian protection. ADAS use cases are typical examples of driver assistance applications.

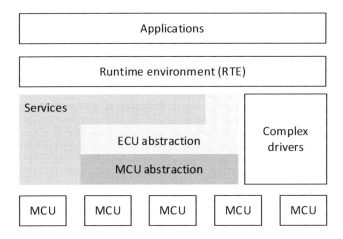

Figure 4.9: AUTOSAR software architecture

AUTOSAR supports very different vehicle system configurations and network topologies. There are communications stacks for vehicle onboard units and RSUs for CAN, FlexRay or Ethernet and there need to be stacks for IEEE 802.11p / ETSI G5 or LTE and 5G as well. Since AUTOSAR defines a development methodology and software infrastructure, partial aspects of part 6 of ISO 26262, which encloses the requirements for the development of safety-related software are implemented.

Whereas AUTOSAR is clearly targeting vehicle control and ADAS, there is another software activity that has software specifications and standards in the vehicle ecosystem pursuing the development of infotainment applications. GENIVI focuses on middleware providing a software toolbox for a vehicle IVI platform. Function domains, are for example, software management (e.g. SOTA), networks (CAN, Flexray, USB, Wi-Fi, NFC, and Bluetooth, etc.), navigation and location based services, and telephony. Regarding V2X networking and connectivity, there are currently projects in the area of smart device links (a set of protocols and messages that connect applications on a smartphone to a vehicle head unit) and remote vehicle interaction (to provide robust and secure communications between a vehicle and the rest of the world).

Another important software topic strongly related to V2X networking and connectivity is the over-the-air (OTA) firmware and software update for vehicles. The major challenges are: the rising demand for connected vehicle devices, changing government regulations regarding safety and cyber security of the vehicle, and increasing request for advanced navigation, telematics and infotainment. On top of that, vehicle manufacturers are urged to protect vehicle data from remote hacking and malfunctioning, which in turn rises the call for OTA software updates for vehicles as well.

4.5 HAD maps

We are all aware today, that safe, reliable and trustworthy vehicle navigation systems depend heavily on dynamically updated high definition maps. We define a high definition map as in-vehicle data base comprised of a multi-layer data stack where the layers are interlinked. The objective of this dynamic HD map is to provide the localization, to sense vehicle surroundings, to perceive and fuse sensor data, to support reasoning and decision making, as well as to provide input for motion control. The layer 1 static map is comprised of the basic digital cartographic, topological and road facilities data. The quasi-static layer 2 includes planned and forecasted traffic regulations, road work and the weather forecast. The dynamic layer 3 encompasses traffic data including accidents, congestion and local weather. And the highly-dynamic layer 4 is comprised of the path data, surrounding vehicle and pedestrian data and timing of traffic signals.

Maps for navigation and path planning are typically for human consumption. They get updated at least once a year and are cartographed by special surveillance vehicles with human editing. The online vehicle navigation is done by the driver in real-time and is supported by an upfront offline route planning using maps. Evolving from navigation toward automated and autonomous driving, the maps for vehicles are generally for computing and are up to date and cartographed for example by any regular vehicles on the road with automated data processing. Vehicle sensors and vehicle data fuel automatically the map update together with V2X networking and computing. Then, the vehicle navigation gets tightly combined with the strategic, tactical and reactive vehicle path planning with very different time and location constraints.

Highly automated driving (HAD) maps include highly detailed inventories of all stationary physical assets related to streets such as lanes, road edges, shoulders, dividers, traffic signals, signage, paint markings, poles, and all other critical data needed for the direction finding on roadways and intersections by automated and autonomous vehicles. HAD map features are highly accurate (in the cm absolute ranges) in location and time and get updated in real-time via cloud and crowd sourcing. Dynamic live HAD maps deliver data beyond the line of sight for electronic horizon (dynamic eHorizon) predictive awareness to let autonomous vehicles know what lies ahead. The vehicle looks for example beyond 300 m and around the corner with the HAD map model provided.

The path planning (Figure 4.10) for example the vehicle plans the next highway exit, and is not real-time. It delivers the route to a selected destination for an automated or autonomous drive. The appropriateness of street segments are considered for the route calculation using an extended navigation map. In the path planning on lane level, for instance the vehicle has to take the right turn at the intersection, is near real-time. It produces lane change advices according to the vehicle's driving lane and lane traffic situation for upcoming manoeuvres along the vehicle route. This planning uses the HAD map with detailed lane data and lane accurate positioning. Finally, there is the path planning at the lane and geometry level (e.g. the vehicle must avoid any accident) which is strictly real-time. It computes the automated or autonomous vehicle trajectories based on lane change advice and considers surrounding traffic and senses surrounding environment characteristics. This planning uses sensor and environment object data from video camera, ultrasonic sound, and radar and LIDAR sensors.

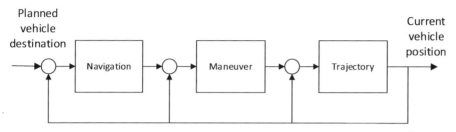

Figure 4.10: Multi-stage control for vehicle path planning

The vehicle with V2X the HAD map data get continuously updated with data from other vehicles, pedestrians and other ecosystem stakeholders. The vehicle acts as another sensor with adaptable range and resolution. The sheer amount of data as well as communication performance required and their scalability for instance between urban and rural scenarios is currently an issue. The implementation of deep learning and artificial intelligence requires the application of a continuous feedback loop comprised of data gathering (shadow mode), DNN training, inference and SOTA update. During shadow mode, the vehicle records large amounts of vehicle data and uploads them to servers in the cloud. New inference models, fail safe and fall back instructions, firmware and other software updates, happen with SOTA.

ADAS use cases which benefit from the electronic horizon and HAD map system architecture (Figure 4.11) for fully automated and autonomous driving vehicles SAE level 4 uses cases like: traffic jam assist, highway pilot or automated valet parking. SAE level 3 uses cases for active driver assistant systems are: adaptive cruise control, forward collision warning or lane keep assist. Lower SAE level use cases for efficient driving or passive driver assistant systems are obviously supported as well. Examples are eco drive assist, range control assist, intersection warning, traffic sign assist, curve speed warning, predictive curve light or night vision.

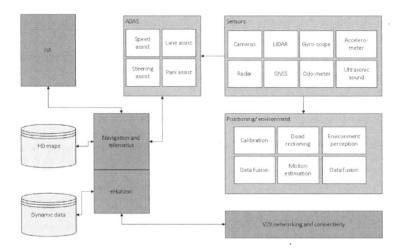

Figure 4.11: HAD architecture with electronic horizon

The electronic horizon provides other vehicle ECUs a continuous prediction of the upcoming street network using standardized data exchange protocols between navigation and telematics and HAD maps. It integrates map matched localization, most probable path estimation, static map attributes including curvature, slopes, speed limits, road class, etc. and dynamic data as of route, traffic data, hazard warnings, road construction data, weather and so on. A perception layer, aims to detect the conditions of the environment surrounding the vehicle, for instance, by identifying the appropriate lane and the presence of obstacles on the track. A reference generation layer, which is based on the inputs from the perception layer, provides the reference signals as kind of a reference trajectory to be followed by the vehicle. And a control layer defines the commands required for ensuring the tracking performance of the reference trajectory. These commands are usually expressed in terms of reference steering angles and traction or braking torques and are sent to the vehicle control ECUs.

The advancing map-enhanced driver assistance systems (ADASIS) consortium uses the standard protocol for exchanging electronic horizon. The electronic horizon reconstruction ensures electronic horizon protocol integrity as it runs on an ECUs domain controller and is therefore subject to ASIL allocations and functional safety. ADASIS version 2 is comprised of the standard and ADAS map data for advanced driver assistance use cases and is designed for data broadcast on the CAN bus. The most probable path and tree length is up to 8190 meters with an attribute resolution of meters. ADASIS version 3 consists of HAD map data for automated and autonomous driving up to SAE level 5. It supports broadband communication as of Ethernet and TCP/IP, with bi-directional communication support for example in a P2P mode. The most probable path and tree length is up to 43000 kilometers with an attribute resolution of centimeters. This enables automated and autonomous driving HAD features like an extended lane model (road and lane geometry, road and lane width, lane marking, lane connectivity), highly accurate junction model (lane merges and markings, splits, stop lines), a speed profile related to specific road and lane segments (real-time speed profiles, speed profile histories), vehicle within surrounding (unique ID, position on the link and lane path, speed, vehicle status) and parking area models (geometry).

4.6 Functional safety

Automated and autonomous driving vehicles will never make it out of test trials unless issues regarding privacy, functional safety, and security have been addressed. The term privacy is used here as the person's freedom from interference or intrusion by others and in particular the ability to determine what data about him or her are shared. We use the term functional safety for the correct system functioning of the vehicle and the protection of the driver and passengers in the vehicle, mainly to ensure avoidance of vehicle or traffic accidents. The term security refers to protective digital privacy measures where the integrity, confidentiality, and availability of an individuals' data are guaranteed.

Functional safety is one of the hottest topics as semiconductor manufacturer's wireless solutions for automated and autonomous vehicles become available now. Automotive systems must be compliant to ISO 26262 and its associated automotive safety integrity levels (ASILs) B through D. ASIL B nominal safety use cases are for example lane assist, park assist, speedometer, rear camera or the ones where it's sufficient to make the driver aware if the system is not working. ASIL D relates to critical safety use cases like braking, steering, acceleration, chassis control, air bag, seat belt tension or the ones where the driver relies on the systems to function correctly all the time. Standards like IEC 61508 cover industrial functional safety and are important to look at as well, since it applies to electrical, electronic and programmable electronic systems. These standards mandate robust error detection and mitigation to ensure a vehicle does not run into a risk of hazards caused by system malfunction and continues to operate safely even after a component failure.

ISO 26262 must be implemented from intellectual property design to system implementation (accompanied by comprehensive documentation for the applicable requirements). The vehicle systems must be capable of dealing with systematic faults (hardware errata, software bugs, incorrect specifications, incomplete requirements) as well as random faults (hardware failures, memory errors, permanent, transient or latent errors). There are different options to do so. First there is the option to implement diverse systems and choose one out of them (dual, triple units etc., different implementations, random and systematic choice). Second redundant hardware blocks are used by adding checks and third, there is redundant execution running functions multiple times and double checking the results.

Users expect privacy and security in their cars. Consequently, the collection, processing, and linking of data have to be in accordance to the laws of privacy. At present, a lot of personalized data is already collected via navigation systems, smartphones, or during vehicle maintenance. Automated vehicles are capable of recording and providing large amounts of data that might assist crash investigations and accident reconstructions. Such data is of high relevance for improving active safety systems and system reliability but also for resolving liability issues. Existing accident data bases such as the GIDAS project (German in-depth accident study) are updated and extended continuously by automated driving information (SAE level classification of involved vehicles, driver/automated mode etc.).

Furthermore, cyber security as a vulnerability to hacking has to be considered in order to avoid that the vehicle or driver lose control over the vehicle due to intrusion and theft, malicious hacking or unauthorized updates for instance. Vehicles offer multiple vulnerabilities for attacks in particular with the implementation of wireless networking and connectivity. Attacks can happen to the IVI, the navigation and telematics, the vehicle control, the connectivity or the ADAS system. These attacks can happen externally via wireless (Wi-Fi, Bluetooth, NFC, RKE, LTE, 5G) links or from inside the vehicle via IVI and on-board diagnostics interfaces.

Probable attack points are: the remote anti-theft system, tire pressure monitoring system (TPMS), remote keyless entry and start, Bluetooth, the radio data system, the navigation and telematics and connectivity system. Some possible defences against attacks are the minimization of attack points, communications protocol message injection mitigation, communication protocol message cryptography, vehicle network architecture changes or clock based intrusion detection. In addition to the cyber-attacks into the vehicle, there are also attacks against the supporting cloud infrastructure. If the vehicle ecosystem infrastructure gets compromised, the interfaces to the vehicle will get abused. Cloud based wireless connected vehicle services are likely to become most attractive targets for hacking. In particular data on vehicle capabilities, status, location, route and driver and passenger's data need to be protected from cyber-threats.

Vehicle manufacturers and their suppliers are in agreement that V2I and V2V networking and connectivity protocols have to be developed with security embedded along the entire development phase. That means for example that all units connected to the vehicle shall be protected byensuring the units and communications are intrinsically secure, incorporating secure coding and encrypted communications and data privacy is safeguarded, not only by normative expectations such as by law, but already on the vehicle system level. For instance, each connected vehicle unit can only communicate data to the units that are absolutely relevant for its functionality. Furthermore, access is only granted by the units having the corresponding access rights.

One of the most challenging issues are attacks on the software, where hackers exploit errors like software bugs, configuration or specification errors. Plentiful new errors are reported every year in all operating systems including Linux, Windows and others. Operating systems and communications drivers and applications cannot be directly secured and need to be sandboxed in some way. Security by design is a must and is achieved by using a formally proven kernel for protecting the entry points like the IVI, navigation and telematics, vehicle control or connectivity system.

Finally, let's have a look at a security technology which might have the capability to disrupt the way security is implemented in connected vehicles, *blockchain*. But not only that, blockchain could disrupt how vehicles are made, how they are used and how they are maintained. A blockchain is considered as a register of distributed records in batches or blocks. These blocks are securely linked together in a virtual space where each valid block is linked with a time-stamp transaction and linked with the previous block. In the implementation of a secure telematics platform using a blockchain, the platform is capable of authenticating telematics data from the vehicle and the access along with authorization of all vehicle ecosystem stakeholders linked with it. Block chain driven crypto-currency may possibly empower vehicles with a repository which could facilitate the purchase of mobile services like updates, parking fees, toll fees and so on. Blockchain gives the vehicle a digital identity and keeps drivers' and passenger's privacy at the same time.

4.7 Conclusions

There are sensor, computing, communications, software and high-definition maps technologies which have an impact on V2X networking and connectivity. To achieve fully automated and autonomous self-driving vehicle capabilities, SAE level 4 or5, vehicle manufactures must equip vehicles with an array of sensors that can identify pedestrians, road markings, traffic signs, other vehicles, and miscellaneous objects both day and night as well as under all weather conditions that human drivers are typically able to navigate. Generating 3D environment models and range estimation using only video cameras is possible, but requires more complex processing of synchronized stereo images calculating structure from motion. LIDAR is more precise and offers edge detection, which increases processing efficiency by reducing the output data to only regions of interest (ROIs). Combined, the different views of a vehicle's environment made by ultrasonic sound, video cameras, radar and LIDAR, enable range and resolution far exceeding those of human vision. The sensor outputs get integrated into a three dimensional 360-degree environmental model running advanced sensor fusion algorithms on high-performance automated and autonomous driving computing platforms.

Ultrasonic sound, video, radar or LIDAR sensor technology doesn't rely on V2X networking and connectivity for their implementation in vehicles. On the other hand, assuming functional safety gets achieved for V2X networking and connectivity there is a feasibility of high throughput together with bounded latency and high reliability for vehicle sensors in certain use cases. The sharing of full raw sensor data beyond nearest vehicle neighbors in high dynamic traffic scenarios is not feasible, but also most likely not needed since it includes too much data to be sent and contains highly dynamic data which quickly becomes irrelevant with distance. There is the option to use 5G communications technology for the exchange of full sensor data and sharing of full 3D images requiring a very high data throughput. Then the requirements on V2X networking and connectivity are qualitatively set to strictly bounded delay, maximum reliability inside delay bounds and achievable throughput. If it is necessary to ensure high sensor data rates with simultaneously high spatial 3D resolutions, sensor networking and connectivity becomes challenging.

The evolution of powerful central computing platforms for autonomous and automated driving is required due to the sheer amount of data to be processed and the scalability and flexibility to be implemented. The development of computing platforms starts from decentralized ECUs, goes over to central ECUs and evolves toward to computing platforms with server domains and virtualized ECUs.

It is very challenging to develop a wireless V2X networking and connectivity technology able to cope with the extreme complexity of a vehicular network, in terms of mobility, environment dynamics, technology heterogeneity, and to fulfil the often-contradictory requirements of all vehicular use cases. The foremost challenges of V2X are the high vehicle mobility and the high variability of the surroundings in which vehicles run. A huge amount of different configurations are possible, ranging from highways with relative inter-vehicle velocity of up to 300 km/h and a comparatively low spatial density, to urban city crossings where relative inter-vehicle velocity is on the order of lower tens of km/h and spatial density is extremely high.

A connected vehicle with V2X networking and connectivity is expected to have all the functionalities that one expects from their smartphones. But V2X networking and connectivity is not synonymous to autonomous or automated driving. An autonomous or automated vehicle does not require per se V2X. But we do think that V2X is a sensor extension for autonomous and automated vehicles to improve situation awareness, to provide redundancy when sensors fail, to update in vehicle databases and firmware and software over-the-air (FOTA, SOTA) and to enable, for example, telemetry. V2X networking and connectivity helps to improve situational awareness by providing redundancy if sensors fail, resolving traffic bottlenecks and reducing road congestion. Autonomous vehicles have many sensors. They are sufficient alone to make the vehicle moving around and V2X communications is not needed.

The V2X networking and connectivity core standards for DSRC are SAE J2736 and IEEE 1609 and 802.11 in the United States. In Europe these are ETSI TS 103 175, 102 687, 102 724, 102 941, 103 097, 102 539, EN 302 663, and 302 636 and for the CEN/ISO TS 19 321 and 19091, which were published until 2014. Regarding automated and autonomous driving, both C-ITS and DSRC have to be extended. What is lacking currently is a functional safety concept with fail-safe functions.

The vehicle software focus is on firmware now. V2X networking and connectivity introduces new challenges to software like software applications (apps), real-time changes over the air and multiple source software. The broadly implemented AUTOSAR and GENIVI compliant V2X stacks need to be extended for upcoming V2X networking and connectivity standards like LTE and 5G. And V2X networking and connectivity becomes another component of a mobile firmware-over-the-air (FOTA) or software-over-the-air (SOTA) system, including to the packager, the vehicle inventory, and the distribution server in the cloud, the vehicle OTA client and the protocol and reporting.

In the HAD map use cases, vehicle navigation system supports the automated or autonomous driving with geo fencing, calculation of HAD routes, generation of HAD manoeuvre advice, delivery of smart safe state locations and synchronizes the navigation user interface with the HAD map status. The electronic horizon of HAD maps connect the worlds of navigation and telematics with highly automated and autonomous driving. Challenges are ahead with the collection and exchange of data, the storage and processing of these data and the creation of precise up-to-date real-time HAD maps. In the case of live dynamic HAD maps the required V2X data throughput per vehicle in uplink and downlink depends on the specific use case. High data throughput use cases, for instance, are related to HAD maps with occupancy grid and full LIDAR, video camera or radar sensor images. Moderate data throughput use cases are the sharing of planned trajectories and high level coarse traveling decisions. Low data throughput applications are short emergency messages, short messages to coordinate manoeuvres and the periodic broadcast of vehicle status messages.

Mobile security architectures have converged for many years toward a security architecture based on three pillars. First, securing the elements or hardware coprocessors are implemented for the root of trust, cryptography, and transactions. Second, trusted execution environments (TEE) and secure operating systems ensure a well determined environment to run applications and services. And third, there are software hypervisors. TEE and hypervisors need to be significantly reinforced for automated and autonomous vehicles.

But data privacy and security are specific for the vehicle ecosystem since vulnerabilities have to be prevented in any case under economic constraints of implementation. And the combination of vehicle ADAS and control functions with wireless networking and connectivity requires dedicated efforts for providing functional safety. Functional safety for security and privacy standards must be developed for each level of autonomous and automated driving, including in-vehicle computing and communication, inter-vehicle communication, infrastructure and vehicle-infrastructure communication.

ISO 26262 is a must have; in particular for autonomous and automated driving. Vehicle ecosystem hardware and software platforms must inherently deliver security and privacy. It becomes clear that the standard for vehicle functional safety, ISO 26262, needs extensions for connected automated or autonomously driving vehicles. Taking the increasing complexity of automated and autonomous driving vehicles including vehicle networking and connectivity into account, a profound and growing focus on functional safety and security is an absolute inevitability. The challenges and constraints of wireless technologies to be implemented, as well as the potential threats in the vehicle ecosystem must be understood and assessed early and all over the wireless networking and connectivity development life-cycle.

The evolution of V2X technologies should look and feel familiar from its introduction and require virtually no new learning for the drivers, the passengers, and the other autonomous and automated vehicle ecosystem stakeholders. We think that mirroring established user interface conventions, building on existing user behaviors and incorporating well-liked smartphone apps and designs are going to be a successful way forward. V2X networking and connectivity has to contribute to stress reduction and to make drivers and passengers feel safe by building trust. So V2X technologies shall deliver only the right data at the right time since when driving less is more. V2X must fit into a simple and consistent visual look and feel, supporting multi-modal cues for individual preferences. And finally, V2X technologies are going to allow users to participate in and shape the predictions vehicles make on their behalf, with challenges to make privacy settings obvious and easy to use, give drivers and passengers more control up front when on boarding new features, and to quickly take back control in an easy way.

The key performance indicators for V2X networking and connectivity are reliability, data throughput, and security, but functional safety becomes a key challenge for the integration of sensor, computing, communication, software and high-definition maps technologies due to the specifics of wireless communication technologies.

References

Altintas, Onur (Ed.) (2016): 2016 IEEE Vehicular Networking Conference (VNC). 8-10 Dec. 2016. Institute of Electrical and Electronics Engineers; IEEE Vehicular Networking Conference; VNC. Piscataway, NJ: IEEE. Available online at http://ieeexplore.ieee.org/servlet/opac?punumber=7822829.

Bojarski, Mariusz; Testa, Davide Del; Dworakowski, Daniel; Firner, Bernhard; Flepp, Beat; Goyal, Prasoon et al.: End to End Learning for Self-Driving Cars. Available online at http://arxiv.org/pdf/1604.07316v1

Fossen, Thor I.; Pettersen, K. Y.; Nijmeijer, H. (2017): Sensing and control for autonomous vehicles. Applications to land, water and air vehicles / Thor I. Fossen, Kristin Y. Pettersen, Henk Nijmeijer, editors. Cham, Switzerland: Springer (Lecture notes in control and information sciences, 0170-8643, volume 474). Available online at http://link.springer.com/ BLDSS

2011 International Joint Conference on Neural Networks (IJCNN 2011 - San Jose). San Jose, CA, USA.

2016 IEEE Intelligent Vehicles Symposium (IV). Gotenburg, Sweden.

2016 IEEE Vehicular Networking Conference (VNC). Columbus, OH, USA.

2017 IEEE 47th International Symposium on Multiple-Valued Logic (ISMVL). Novi Sad, Serbia.

ABI Research (2016): Connected Automotive Tier One Suppliers. ABI Research.

Accenture (1/10/2014): Connected Vehicle Survey Global.

China Unicom (2017): Edge Computing Technology. China Unicom.

ETSI ITS (2017): Certificate Policy for Deployment and Operation of European Cooperative Intelligent Transport Systems (C-ITS). ETSI ITS.

European Automotive and Telecom Alliance (5/19/2017): Codecs at Brussels.

Infineon (2016): Your path to robust and reliable in-vehicle networking. Infineon's automotive networking solutions. Infineon.

OVERSEE (2011): Use Case Identification. OVERSEE.

Renesas Electronics Corporation (2016): Renesas Automotive. Renesas Electronics Corporation.

U.S. Department of Transportation (2016): Privacy Impact Assessment for V2V NPRM. U.S. Department of Transportation.

Altintas, Onur (11/13/2016): A Communication-centric Look at Automated Driving.

Amditis, Angelos (7/25/2016): Automated Driving - the present and beyond.

Baldessari, Roberto (4/4/2017): Trends and Challenges in AI and IoT for Connected Automated Driving.

Bansal, Gaurav (3/27/2015): The Role and Design of Communications for Automated Driving.

Bolignano, Dominique (4/4/2017): In-vehicle technology enabler.

Bonte, Dominique; Hodgson, James; Menting; Michela (8/11/2015): SDC, ADAS Sensor Fusion, and Machine Vision.

Chen, Shitao; Zhang, Songyi; Shang, Jinghao; Chen, Badong; Zheng, Nanning (2017): Brain-inspired Cognitive Model with Attention for Self-Driving Cars. In IEEE Trans. Cogn. Dev. Syst., p. 1. DOI: 10.1109/TCDS.2017.2717451.

Dimitrakopoulos, George (2017): Current Technologies in Vehicular Communication. Cham: Springer International Publishing; Imprint: Springer.

Eckert, Alfred (11/12/2015): From Assisted to Automated Driving.

Fabozzi, Frank J.; Kothari, Vinod (Eds.) (2008): Introduction to Securitization. Hoboken, NJ, USA: John Wiley & Sons, Inc.

Filippi, Alessio; Klaassen, Marc; Roovers, Raf; Daalderop, Gerardo; Walters, Eckhard; Perez, Clara Otero (10/21/2016): Wireless connectivity in automotive V2X over 802.11p and LTE: a comparison.

Fleming, William J. (2008): New Automotive Sensors—A Review. In IEEE Sensors J. 8 (11), pp. 1900–1921. DOI: 10.1109/JSEN.2008.2006452.

Flore, Dino (2/27/2017): 5G V2X.

Fogue, Manuel; Garrido, Piedad; Martinez, Francisco J.; Cano, Juan-Carlos; Calafate, Carlos T.; Manzoni, Pietro: A Realistic Simulation Framework for Vehicular Networks. In George Riley, Francesco Quaglia, Jan Himmelspach (Eds.): Fifth International Conference on Simulation Tools and Techniques. Desenzano del Garda, Italy.

Förster, David (2017): Verifiable Privacy Protection for Vehicular Communication Systems. Wiesbaden: Springer Fachmedien Wiesbaden.

Försterlin, Fank (3/4/2017): Automated Driving.

Gage, Tom (2017): Improving vehicle cybersecurity. ATIS.

Gleave, Steer Davies; Frisoni, Roberta; Dionori, Francesco; Vollath, Christoph (2014): Technology options for the European electronic toll service. European Commission.

Gunther, Hendrik-Jorn; Mennenga, Bjorn; Trauer, Oliver; Riebl, Raphael; Wolf, Lars: Realizing collective perception in a vehicle. In: 2016 IEEE Vehicular Networking Conference (VNC). Columbus, OH, USA, pp. 1–8.

Herrtwich, Ralf (4/4/2017): What's in a Map?

ILIE, Irina (2016): ACEA Strategy Paper on Connectivity. ACEA.

Kandel, Paul; Mackey, Matt (5/25/2016): The road to autonomous driving.

Kenney, John B. (2011): Dedicated Short-Range Communications (DSRC) Standards in the United States. In Proc. IEEE 99 (7), pp. 1162–1182. DOI: 10.1109/JPROC.2011.2132790.

Kisacanin, Branislav: Deep Learning for Autonomous Vehicles. In : 2017 IEEE 47th International Symposium on Multiple-Valued Logic (ISMVL). Novi Sad, Serbia, p. 142.

Lakrintis, Angelos (2017): Mentor Graphics DRS360: Utilizing Raw Data for Level 5 Autonomous Driving. Strategy Analytics.

Lanctot, Roger (9/26/2016): HERE Open Location Platform Offers New Data Sharing Model and Monetization Opportunities for the Auto Industry.

Lauxmann, Ralph (4/4/2017): Enhanced vehicle automation functions to improve road safety.

Locks, Olaf; Winkler, Gerd (3/8/2017): Future Vehicle System Architecture ASAM General Assembly.

Lunt, Martin (3/13/2017): AUTOSAR Adaptive Platform.

Magney, Phil (5/25/2017): The Enablement of Automated Driving.

Mak, Kevin (2017): Intel's New Autonomous Vehicle Center: The Importance of Cross-Domain Development. Strategy Analytics.

Mak, Kevin (5/372017): ARM Mali-C17: Image Signal Processor Targeting Automotive Multi-Camera Systems. Strategy Analytics.

Mak, Kevin: Intel's New Autonomous Vehicle Center: The Importance of Cross-Domain Development.

Mariani, Riccardo (5/25/2017): Applying ISO 26262 to ADAS and automated driving.

Meixner, Gerrit; Müller, Christian (2017): Automotive user interfaces. New York NY: Springer Berlin Heidelberg.

Misener, Jim (7/27/2015): Applying Lessons Learned to V2X Communications for China.

Mueller, Juergen (6/9/2017): TS-ITS100 RF Conformance Test System for 802.11p.

Ozbilgin, Guchan; Ozguner, Umit; Altintas, Onur; Kremo, Haris; Maroli, John: Evaluating the requirements of communicating vehicles in collaborative automated driving. In : 2016 IEEE Intelligent Vehicles Symposium (IV). Gotenburg, Sweden, pp. 1066–1071.

Pruisken, Stefan (9/19/2017): Highly accurate navigation in the age of automated driving.

R. Saracco; S. Péan (3/21/2017): 5G the enabler of connected and autonomous vehicles.

Redfern, Richard (2015): Key Performance Indicators for Intelligent Transport Systems. AECOM Limited.

Riley, George; Quaglia, Francesco; Himmelspach, Jan (Eds.): Fifth International Conference on Simulation Tools and Techniques. Desenzano del Garda, Italy.

Schaub, Norbert; Becker, Jan (6/27/2017-6/28/2017): Context-embedded vehicle technologies.

Schwarz, Stefan; Philosof, Tal; Rupp, Markus (2017): Signal Processing Challenges in Cellular-Assisted Vehicular Communications. Efforts and developments within 3GPP LTE and beyond. In IEEE Signal Process. Mag. 34 (2), pp. 47–59. DOI: 10.1109/MSP.2016.2637938.

Sermanet, Pierre; LeCun, Yann: Traffic sign recognition with multi-scale Convolutional Networks. In : 2011 International Joint Conference on Neural Networks (IJCNN 2011 - San Jose). San Jose, CA, USA, pp. 2809–2813.

Shulman; Deering (2007): Vehicle safety communications in the U.S. Ford Motor Company; General Motors Corporation.

Steiger, Eckard (4/4/2017): In-Vehicle Technology Enabler for CAD.

Subash, Scindia; Lunt, Martin (3/8/2017): E/E architecture in a connected world.

Takada, Hiroaki (6/8/2012): Introduction to Automotive Embedded Systems.

Torsten Schütze; Rohde & Schwarz SIT GmbH (2008): Introduction. In Frank J. Fabozzi, Vinod Kothari (Eds.): Introduction to Securitization. Hoboken, NJ, USA: John Wiley & Sons, Inc, pp. 1–12.

Tuohy, Shane; Glavin, Martin; Hughes, Ciaran; Jones, Edward; Trivedi, Mohan; Kilmartin, Liam (2015): Intra-Vehicle Networks. A Review. In IEEE Trans. Intell. Transport. Syst. 16 (2), pp. 534–545. DOI: 10.1109/TITS.2014.2320605.

Vantomme, Joost: Frequency bands for V2X. ACEA. 12/1972016.

Verlekar, Vibhav (8/30/2017): Rapid Evolution of 'The Connected Car' Changing to the Digital Lane.

Yang, Bo; Zhou, Lipu; Deng, Zhidong (2013): Artificial cognitive model inspired by the color vision mechanism of the human brain. In Tsinghua science and technology 18, pp. 51–56.

Chapter 5
V2X networking and connectivity

In this chapter, we look at vehicular communication technologies—with 3GPP LTE C-V2X and IEEE 802.11p-based DSRC (U.S.) and ITS-G5 (Europe) being the main candidates—their respective feature sets, required modifications to existing modem technologies, and specific deployment challenges, such as high-speed operation and high node density, and so on. We further discuss issues of existing V2X protocols for networking vehicles, which are for IEEE 802.11p the poor scalability, lack of QoS guarantee, intermittent V2I connectivity, and high cost for coverage. For LTE V2X, penalties seen are the infancy at 3GPP which most likely is going to evolve and adapt existing LTE ProSe protocol for V2X, hard to have significant improvement for a broadcast-oriented protocol for public safety mainly. LTE V2X is currently designed mainly for CAM (cooperative awareness messages) and DENM (decentralized environmental notification messages) with 100ms latency, greater than 80% reliability at unknown density.

5.1 IEEE 802.11p-based DSRC and ITS-G5

In this section, we consider the IEEE 802.11p-based Dedicated Short-Range Communications (DSRC) solution as it is applied in the U.S., as well as the European ITS-G5 system, which employs a European profile of the physical and medium access control sub-layer using IEEE 802.11 as the base standard. According to the U.S. National Connected Vehicle Field Infrastructure Footprint Analysis Final Report, a connected vehicle infrastructure deployment is expected to leverage technologies that comprise cellular and Wi-Fi as well as DSRC, and would typically include:

- Roadside communications equipment (for DSRC or other wireless services) together with enclosures, mountings, power, and network backhaul
- Traffic signal controller interfaces for applications that require signal phase and timing (SPaT) data
- Systems and processes required to support management of security credentials and to ensure a trusted network
- Mapping services that provide highly detailed roadway geometries, signage, and asset locations for the various connected vehicle applications
- Positioning services for resolving vehicle locations to high accuracy and precision
- Data servers for collecting and processing data provided by vehicles and for distributing information, advisories, and alerts to users

DOI 10.1515/9781501507243-005

Such a deployment would serve the public good for the U.S. context as summarized in the Footprint Report:

- The number and severity of highway crashes will be dramatically reduced when vehicles can sense and communicate the events and hazards around them.
- Mobility will be improved when drivers, transit riders, and freight managers have access to substantially more updated, accurate, and comprehensive information on travel conditions and options, and when system operators—including roadway agencies, public transportation providers, and port and terminal operators—have actionable information and the tools to affect the performance of the transportation system in real-time.
- The environmental impacts of vehicles and travel can be reduced when travellers can make informed decisions about the best available modes and routes, and when vehicles can communicate with the infrastructure to enhance fuel efficiency by avoiding unnecessary stops and slow-downs.

The 2012 FHWA report on Crash Data Analysis for Vehicle-to-Infrastructure Communications for Safety Applications summarizes expected benefits stemming from the deployment of vehicle-to-infrastructure (V2I) applications and related safety improvements. It provides estimates of the frequency and cost of crashes involving precrash scenarios addressed by V2I applications. The report concludes that "currently identified V2I safety applications could potentially target approximately 2.3 million crashes and $202 billion in costs," assuming the applications are 100% effective in eliminating those crashes and deployed everywhere in the United States.

The overall connected vehicle system diagram suggested by the report is indicated in Figure 5.1.

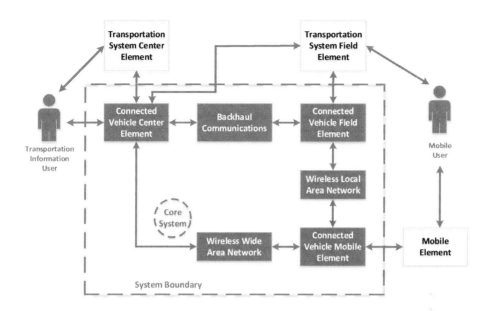

Figure 5.1: Connected vehicle system diagram.

Concerning a specific vehicle, certain components need to be included in a vehicle, as outlined Figure 5.2.

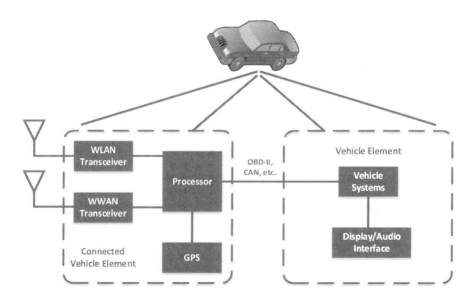

Figure 5.2: Embedded connected vehicle terminal example.

For the full IEEE 802.11p-based DSRC deployment, the U.S. Government Accountability Office identified a number of challenges, particularly for V2I communications, which are summarized below:

- Spectrum sharing—allowing unlicensed devices to operate in the 5.9 GHz band along with V2I technologies
- Non-mandatory, voluntary deployment of V2I technologies among states and localities
- Limitations at the federal level for V2I deployment in terms of funding or providing technical resources to state/local transportation operators
- Limitations at the state/local level for V2I deployment (funding, personnel, or other areas)
- Ensuring privacy under a system that involves the sharing of data among vehicles and government infrastructure
- Data security measures are in place to secure the data that is collected at the test bed
- Ensuring data security in deploying V2I technologies
- Standardization
- Human factors in the deployment of V2I technologies
- Liability issues, in terms of uncertainty related to legal responsibility for vehicle crashes using V2I technology

The expected timeline for connected vehicle infrastructure deployment milestones from the Footprint Analysis Report is summarized Figure 5.3.

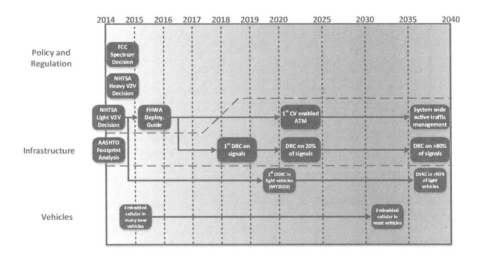

Figure 5.3: Estimated connected vehicle infrastructure deployment milestones.

The expected deployment timelines presented for the various application types from the Footprint Analysis report are summarized Figure 5.4.

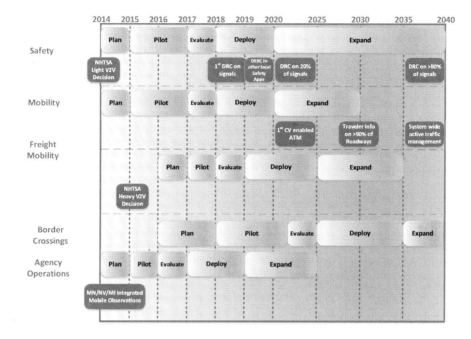

Figure 5.4 Deployment timelines by application type.

In order to design an IEEE 802.11p-based DSRC modem product, Amendment 6: Wireless Access in Vehicular Environments of the IEEE 802.11 standard, the following needs to be considered. While key principles of the IEEE 802.11 baseline (as used for Wi-Fi off-the-shelf products) remain unchanged, a key difference with commercial Wi-Fi products consists in the modified frequency band (5.855 GHz to 5.925 GHz) and the definition of a 10 MHz channel bandwidth. From a modem design perspective, the 10 MHz channel bandwidth choice is challenging because available commercial Wi-Fi modems typically build on a 20 MHz bandwidth and above. However, there is IEEE 802.11j, a specific extension of the IEEE 802.11 standard for the needs of the Japanese market. IEEE 802.11j includes a 10 MHz mode and thus it may be a valid option to build IEEE 802.11p-based DSRC modems with existing IEEE 802.11j designs as a baseline.

Note that in Europe, the term "ITS-G5" has been introduced and is used instead of the term "DSRC." ETSI ES 202 663. It specifies the European profile of the physical and medium access control sub-layer of 5 GHz intelligent transport systems (ITS) using IEEE 802.11 as the base standard. It covers the following frequency ranges:

- ITS-G5A: Operation of ITS-G5 in European ITS frequency bands dedicated to ITS for safety-related applications in the frequency range 5,875 GHz to 5,905 GHz
- ITS-G5B: Operation in European ITS frequency bands dedicated to ITS non-safety applications in the frequency range 5,855 GHz to 5,875 GHz
- ITS-G5C: Operation of ITS applications in the frequency range 5,470 GHz to 5,725 GHz

ETSI ES 202 663 covers the following IEEE 802.11 services:
- Spectrum management services (DFS, uniform spreading) for ITS-G5C
- Transmit power control
- Traffic differentiation and QoS support
- Selected MAC data services: DCF, EDCA, fragmentation/de-fragmentation (the latter only for ITS-G5C)
- Selected MAC control services: ACK, RTS, CTS
- Selected MAC management services: Selected action frames (spectrum management action frames)
- OFDM PHY

The profile excludes the following IEEE 802.11 features:
- Association services
- Access to control and data confidentiality services
- Higher-layer timer synchronization
- Selected MAC data services, i.e. PCF, HCF HCCA
- Selected MAC control services, i.e. PS-Poll, CF-End, CF-End + CF-Ack, Block Ack Request/Block Ack
- Selected MAC management services, i.e. beacon, ATIM, disassociation, association request/response
- Re-association request/response, probe request/response, authentication, de-authentication, selected action
- Measurement request/report
- Power-management services

A key difference from the U.S. DSRC design, ITS-G5 introduces a requirement on "Decentralized Congestion Control (DCC)". This requirement is a mandatory component of ITS-G5 stations operating in ITS-G5A and ITS-G5B frequency bands to maintain network stability, throughput efficiency, and fair resource allocation to ITS-G5 stations. DCC requires components on several layers of the protocol stack, and these components jointly work together to fulfil the following operational requirements:
- Provide fair allocation of resources and fair channel access among all ITS stations in the same communication zone
- Keep channel load caused by periodic messages below pre-defined thresholds

- Reserve communication resources for the dissemination of event-driven, high-priority messages
- Provide fast adoption to a changing environment (busy/free radio channel)
- Keep oscillations in the control loops within well-defined limits
- Comply with specific system requirements, e.g. reliability

Further specificity of the European ITS-G5 system is in the introduction of a "facilities layer" in the protocol stack. The facilities layer is integrated into the ITS-G5 reference architecture as defined in the ITS station reference architecture.

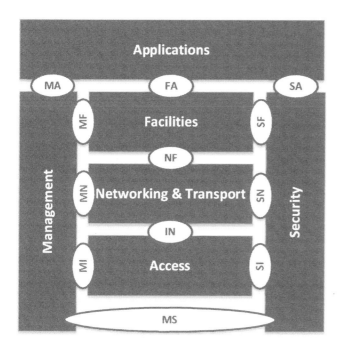

Figure 5.5: ITS station reference architecture/ITS-S Host.

The facilities layer contains functionality from the OSI application layer, the OSI presentation layer (ASN.1 encoding and decoding, and encryption), and the OSI session layer (inter-host communication) with amendments dedicated to ITS:

- Application support (station positioning, service management, message management, mobile station dynamics, security access, etc.)
- Information support (Local Dynamic Map (LDM) database, data presentation, location referencing, station type/capabilities, etc.)
- Communication support (addressing mode, mobility management, GeoNet support, session support, etc.)

- Session support
- Facilities layer management
- Several interfaces

The facilities layer provides support to ITS applications, which can share generic functions and data according to their respective functional and operational requirements. A non-exhaustive presentation of generic functions and data is provided in section 6.3.3 of the ITS station reference architecture. In particular, it uses two central message types:
- Cooperative Awareness Messages (CAM)
  - o Periodic time-triggered position messages
  - o 1–10 Hz, packet length including security up to 800 bytes
- Decentralized Environmental Notification Messages (DENM)
  - o Event-driven hazard warnings

The Local Dynamic Map (LDM) is a database for storing and maintaining data. It provides mechanisms to receive dynamic data from other ITS stations through CAM and DENM. Applications retrieve relevant data from the LDM.
Finally, a basic set of applications are defined in the Local Dynamic Map report. Examples include:
- Road safety (driving assistance): Emergency vehicle, slow vehicle, wrong way driving, traffic conditions, roadwork, etc.
- Traffic efficiency: Speed limits notification, enhanced route guidance, etc.

5.2 LTE and 5G NR V2X

As illustrated below, fifth generation (5G) communication services will provide substantial improvements over 4G services and will include features that are designed for new markets, so-called "verticals," instead of addressing the needs of the classical smartphone user. The addition of new markets is in fact a key component of a viable 5G business model. Today's existing subscription pricing for 4G services is expected to be already in the upper price range that is acceptable to customers, so a substantial revenue increase is unlikely. Consequently, smartphone applications alone are a weak argument for justifying the development and deployment of the 5G ecosystem. There is indeed greater potential in adding new vertical industries to the 5G ecosystem and increasing the revenue potential correspondingly as outlined in "5G Automotive Vision, The 5G Infrastructure Public Private Partnership Project, European Commission."

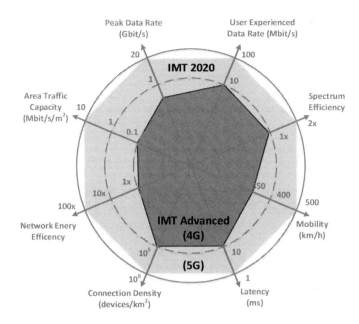

Figure 5.6: 5G capabilities compared to 4G.

Vehicle-to-everything (V2X) communication is one of the key vertical applications and of key interest to mobile network operators to extend their revenue base. Other potential verticals may include industrial automation, health services, program-making and special events (PMSE), and so on. 3GPP has indeed fully committed to addressing the V2X opportunities as a high priority in the context of its three major 5G directions:

–   eMBB (enhanced Mobile Broadband)
–   mMTC (massive Machine-Type Communications)
–   URLLC (Ultra-Reliable and Low Latency Communications)

The corresponding 5G vision and the positioning of V2X applications are illustrated in Figure 5.7 which was inspired by "The path to 5G: as much evolution as revolution - 3GPP".

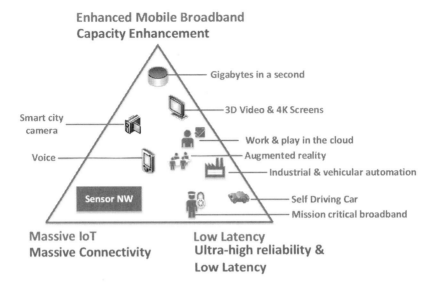

Figure 5.7: 5G capabilities.

For V2X applications, 3GPP has established the following solid C-V2X evolution path to support advanced ITS services. Three development phases are defined, and a fourth development phase is in the planning stage:

- C-V2X Phase 1 – Initial version of C-V2X design based on LTE R14 technology
- C-V2X Phase 2 – Enhanced version of C-V2X design based on LTE R15 technology
- C-V2X Phase 3 – V2X design based on the NR technology (possibly also LTE) R16+
- C-V2X Phase 4 – Further enhancements of C-V2X design based on the NR technology R17+

The timeline for phases two to four, in alignment with 3GPP Releases 15 to 17 are shown in Figure 5.8.

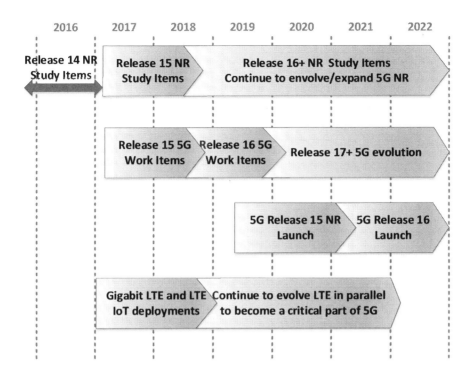

Figure 5.8: 5G timeline.

5.3 C-V2X Phase 1 where 3GPP works on LTE Vehicular Services

3GPP has been working on vehicular communication requirements right from the beginning of LTE Rel. 8. The system had indeed been optimized for low mobile speed from 0 to 15 km/h, but a higher mobile speed between 15 and 120 km/h is being supported with high performance. Mobility across the cellular network is maintained at speeds from 120 km/h to 350 km/h (or even up to 500 km/h depending on the frequency band). Also, wide-range coverage is compatible with vehicular applications. Throughput, spectrum efficiency, and mobility targets are being met at up to 5 km cells, and with a slight degradation at up to 30 km cells. Cell ranges up to 100 km are explicitly not to be precluded. Obviously, the entire system design builds on a cellular infrastructure—cellular base stations are maintaining all control and traffic flow management of communication links. Device-to-device features, such as vehicle-to-vehicle communication, did not find sufficient support in the 3GPP community and therefore are not available among the proposed feature set.

IEEE 802.11p took advantage of this weakness in 3PP LTE—the lack of a device-to-device communication mode—and optimized the entire IEEE system design with a focus on this direct communication technology, advocating it as a key requirement for vehicular applications. Key arguments include low-latency requirements, enabling cars to communicate short term, safety-related features to surrounding vehicles, such as emergency braking. It is argued that the communication via network infrastructure will lead to less advantageous latency levels.

As a consequence, 3GPP developed a device-to-device communication mode (Phase 2) specifically optimized for vehicular applications in Rel. 13 and finally published a more extensive set of vehicular features in Rel. 14. The latter is considered to be the first LTE release to offer a comprehensive set of vehicular services. It was developed over a period of two years, from March 2015 to March 2017. As illustrated in Figure 5.9, a corresponding mass deployment is expected by 2023. Obviously, this timeline leads to a head start of roughly eight years for IEEE 802.11p-based DSRC, which was finalized in 2009. It is indeed unclear whether this time advantage will be sufficient to establish IEEE 802.11p-based DSRC as a main solution in the long-term, or whether it will be overtaken by LTE as soon as it is available in the market and will further evolve toward 5G. Some argue that it is hard to abandon a system that has achieved considerable market share (as we currently see it for second generation (2G) communication systems, such as the Global System for Mobile (GSM) Communications, which still captures considerable market share while the industry is currently transitioning from the fourth to the fifth generation of cellular technology). Others believe that an evolution of IEEE 802.11p-based DSRC is unlikely and will be replaced by a constantly evolving 3GPP solution which is transitioning from fourth generation (4G) to fifth generation (5G) solutions.

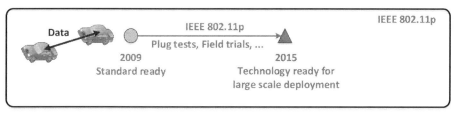

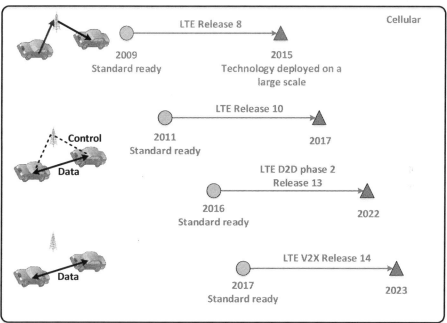

Figure 5.9 IEEE 802.11p and LTE V2X timelines.

The table below presents an overview of the 3GPP Rel. 14 activities on vehicular services, including Study Item (SI, blue color) and Working Item (WI, green color) references. The work is indeed split over a number of 3GPP groups (see the first column of the table below):

— System Architecture Working Group 1 (SA1) defines "service and feature requirements applicable to mobile and fixed communications technology".

— System Architecture Working Group 2 (SA2) provides "definition, evolution, and maintenance of the overall architecture".

— System Architecture Working Group 3 (SA3) "has the overall responsibility for security and privacy in 3GPP systems".

— Core Network and Terminals Working Group 1 (CT1) "is responsible for the 3GPP specifications that define the User Equipment Core network L3 radio protocols and Core network side".

- Core Network and Terminals Working Group 3 (CT3) deals with "Interworking with External Networks" and it "specifies the bearer capabilities for circuit and packet switched data services, and the necessary interworking functions toward both the user equipment in the UMTS PLMN and the terminal equipment in the external network".
- Core Network and Terminals Working Group 4 (CT4) has the "mandate to specify the protocols within the Core Network including specifications describing the protocol requirements".
- Core Network and Terminals Working Group 6 (CT6) is "responsible for the development and maintenance of specifications and associated test specifications for the 3GPP smart-card applications, and the interface with the Mobile Terminal".
- Radio Access Network Working Group 1 (RAN1) defines the "Radio Layer 1 specification".
- Radio Access Network Working Group 2 (RAN2) defines "Radio Layer 2 and Radio Layer 3 RR specifications".
- Radio Access Network Working Group 3 (RAN3) defines "Iub, Iur, and Iu specifications, along with UTRAN O&M requirements".
- Radio Access Network Working Group 4 (RAN4) defines "radio performance and protocol aspects (system) - RF parameters and BS conformance".
- Radio Access Network Working Group 5 (RAN5) defines "mobile terminal conformance testing".

Note that 3GPP does not publish a "Vehicular Communications" specification or similar. Rather, vehicular features are included into the 3GPP Rel. 14 set of specifications. The same is expected to be the case for future releases of 3GPP. The resulting Technical Reports (TR) for Studies and Technical Specifications (TS) for normative work are indicated in Figure 5.10 (for SA1, it is TR22.885, TR22.886, TS22.185, and so on).

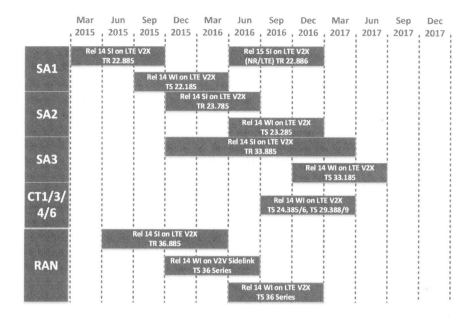

Figure 5.10: 3GPP LTE V2X Planning.

The 3GPP Rel. 14 work on vehicle-to-everything (V2X) communications was mainly driven by road safety applications and related requirements. As summarized in Table 5.1, key requirements relate to high mobility (up to 280 km/h relative speed and in the extreme case, up to 500 km/h) while meeting low latency objectives (time-critical communication enabled by latency < 100ms and even < 20ms in selected use cases, such as pre-crash sensing) and high reliability levels. Targets related to relative vehicle speed have been derived from driver sample response time. Traffic can be either periodic or event-triggered. A typical packet size is ~300 Bytes (up to 1200 Bytes) at a maximum packet rate of ~10 Hz.

Table 5.1: V2X requirements parameter source 3GPP

Operation Scenario	Effective Range	Absolute Speed	Relative Speed	Max. Latency	Min. RX Reliability
#1 (Suburban)	200m	50 km/h	100 km/h	100 ms	90%
#2 (Freeway)	320m	160 km/h	280 km/h	100 ms	90%
#3 (Autobahn)	320m	280 km/h	280 km/h	100 ms	80%
#4 (NLOS/Urban)	100m	50 km/h	100 km/h	100 ms	90%
#5 (Urban intersection)	50m	50 km/h	100 km/h	100 ms	95%

In the following, we will give an overview of basic V2X communication deployment scenarios and related communication types, including vehicle-to-vehicle (V2V), vehicle-to-network (V2N), vehicle-to-infrastructure (V2I), and vehicle-to-person (V2P) communication.

In contrast to IEEE 802.11p-based DSRC, cellular infrastructure will take a special role in 3GPP LTE-based V2X communication. As is shown in Figure 5.11, we differentiate

– Within network coverage (InC): All vehicles are within network coverage
– Partial network coverage: One vehicle is within network coverage and another outside
– Out of network coverage (OoC): All vehicles are outside of network coverage

Note: InC and OoC states are according to a set of (pre-)configured V2X carriers.

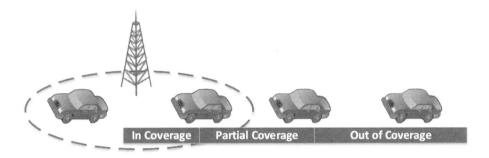

Figure 5.11: Network coverage states

With the cellular infrastructure being present in the communication range of concerned vehicles, it is possible to provide all services through a direct linkage with the infrastructure. 3GPP has defined the so-called "Uu" interface for this purpose (note: the radio interface between the UE and the Node B is called Uu. The radio interface between the UE and the eNodeB is called LTE-Uu). The basic principle is illustrated in Figure 5.12.

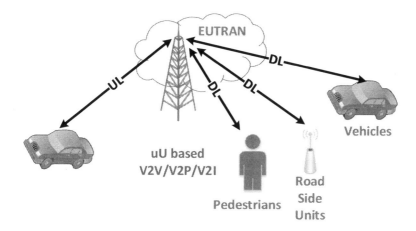

Figure 5.12: Uu interface-based communication

On the other hand, similar to the basic IEEE 802.11p-based DSRC principles, 3GPP has also introduced a so-called sidelink interface (Figure 5.13), a new D2D interface designated as "PC5." The PC5 interface allows direct vehicle-to-vehicle, vehicle-to-person, and vehicle-to-infrastructure communication (such as road side units [RSUs]) as well as integration into the network infrastructure through a combination of LTE-Uu and PC5 communication.

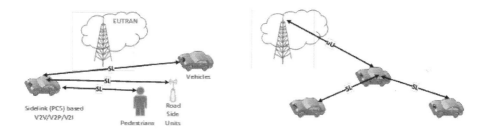

Figure 5.13: Sidelink interface-based communication

In this context, we differentiate between the following V2X communication types:

5.3.1 Vehicle-to-vehicle (V2V) Communication

V2V communication (Figure 5.14) is considered to be one of the main communication modes for vehicular services. It enables direct communication between vehicles, as illustrated below. Typical use cases include emergency braking indications, hazard warnings, and so on.

Figure 5.14: Vehicle-to-vehicle communication

5.3.2 Vehicle-to-pedestrian (V2P, P2V) Communication

V2P/P2V communication services (Figure 5.15) enable vehicles to include the presence and behavior of pedestrians into related decision-making. Related advantages include optimizations targeting pedestrian UE power consumption, and safety-related services with the objective of achieving accident avoidance and thus, a better protection of pedestrians.

Figure 5.15: Vehicle-to-pedestrian, pedestrian-to-vehicle communication

5.3.3 Vehicle-to-infrastructure (V2I, I2V) Communication

V2I/I2V communication services are based on the V2V air-interface communication framework. They enable communication between vehicles and road side units (RSUs). RSUs can be UE- or eNB-type nodes. Note that we differentiate vehicle-to-infrastructure (Figure 5.16) and vehicle-to-network communication—the latter includes a communication link to a (distant) V2X server (typically in the "cloud").

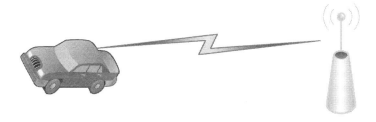

Figure 5.16: Vehicle-to-infrastructure, infrastructure-to-vehicle communication

5.3.4 Vehicle-to-network Communication (V2N, N2V)

V2N/N2V communication services (Figure 5.17) comprise a communication link via RSU/eNB to a (distant) V2X server, as illustrated below. Design assumptions include a sidelink operating at 5.9GHz (ITS band) and a cellular link (Uu) in the 2GHz band.

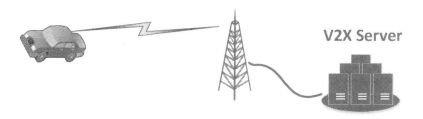

Figure 5.17: Vehicle-to-network, network-to-vehicle communication

While 3GPP Rel. 14 contains a number of vehicular communications features, a specific technical challenge relates to efficient resource allocation. IEEE 802.11p-based DSRC employs a fully distributed resource access scheme based on the carrier sense multiple access/collision avoidance (CSMA/CA) protocol. 3GPP LTE traditionally employs centralized resource allocation. There are many pros and cons for both technologies.

5.3.4.1 CSMA/CA

This distributed protocol does not require any communication overhead for scheduling transmissions, but typically suffers from high packet collision rates and thus low efficiency when the system is operated at a high load levels (i.e., many users are competing for channel access).

5.3.4.2 Centralized scheduling

This scheme is typically employed in cellular communication. The network infrastructure allocates communication resources (time/frequency) for both downlink and uplink. Drawbacks typically include protocol overhead and latency, while the system works efficiently even in a high load scenario.

While sidelink Mode 1 and Mode 2 were defined in 3GPP Rel. 12 for proximity-based services (ProSe) communication, 3GPP Rel. 14 finally defines different resource allocation schemes for Modes 3 and 4. Significant changes have been implemented over previous 3GPP Rel. 12 in order to meet vehicular requirements, including a new UE behavior and physical structure:

5.3.4.3 Mode 3: eNB-Controlled Resource Allocation

The eNB provides specific resources (PSCCH & PSSCH) to be used for vehicle transmission. Sidelink resources can be dynamically or semi-persistently granted.

5.3.4.4 Mode 4: UE-Autonomous Resource Allocation

The eNB (pre)-configures resource pool(s) for PSCCH (control) and PSSCH (shared) channel usage. The UE performs sensing and selects sidelink resources for transmission according to the predefined resource selection procedure.

Sensing and resource selection are optimized for quasi-periodical V2V traffic. Resource reselection is triggered based on a counter (counting number of transmitted TBs). The resource reselection counter is initialized from the range [5, 15] for resource reservation. The basic principle is illustrated in Figure 5.18.

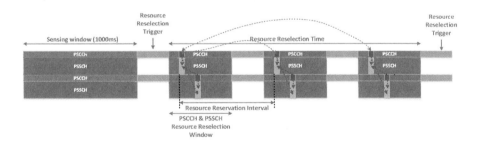

Figure 5.18: Sensing and resource selection for V2V traffic

5.3.5 Synchronization

A further challenge exists in achieving global sidelink V2V synchronization in the time and frequency domains. For this purpose, 3GPP LTE Rel. 14 has defined a synchronization procedure including signals (SLSS) and channels (PSBCH). It is indeed part of the UE capabilities to support SLSS-based transmission and reception.

The following synchronization sources have been defined:
- GNSS can be used as a sync reference within NW coverage
    - o GNSS is the primary reference for OoC scenarios (see more details in sync procedure)
    - o eNBs and UEs propagate timing from eNBs or GNSS can also serve as sync sources
- Multi-Carrier Synchronization (Derived from eNB)
    - o UEs can derive synchronization from a (pre-)configured set(s) of indicated V2X carrier(s) (other than used for SL V2V communication)

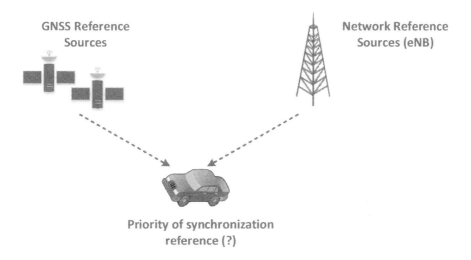

GNSS Reference Sources

Network Reference Sources (eNB)

Priority of synchronization reference (?)

Figure 5.19: GNSS/eNB synchronization references

Furthermore, the following synchronization options are defined:
- GNSS: Directly from GNSS (InC and OoC UE)
    - o UE-GNSS: First hop from GNSS (SLSS-based)
    - o UE-UE-GNSS: Second hop from GNSS (SLSS-based)
- eNB: Directly from eNB
    - o UEe-NB: First hop from eNB (SLSS-based)
    - o UE-UE-eNB: Second hop from eNB (SLSS-based)
- UEISS: Directly from stand-alone UE (independent synchronization source)
    - o UE-UE-ISS: First hop from UEISS

The synchronization source is determined by predefined synchronization source selection (priority) rules. The synchronization procedure precedes V2V transmissions.

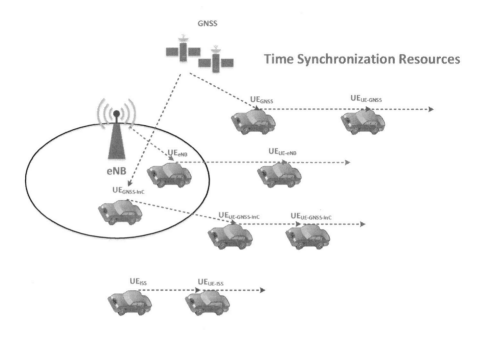

Figure 5.20: Time synchronization resources

5.3.6 Location Determination

Based on GNSS data for exploiting further vehicle location determination services, knowledge about vehicle locations can be exploited for improved overall system efficiency. For example, depending on vehicle location, UEs can be restricted to transmit in an associated resource pool, as illustrated below (the association of resource pools with geo-location zones).

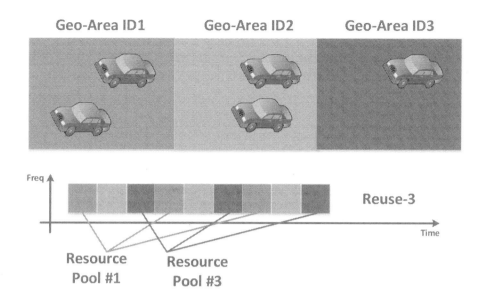

Figure 5.21: Resource pooling

Further benefits of utilizing geo-location include the following:
- Enables spatial reuse in systems with distributed resource selection
- Reduces impact of co-channel interference
- Reduces IBE and near/far problem in vehicular communication scenarios

5.4 C-V2X evolution toward Phase 2 and beyond

3GPP Rel. 15 will define 3GPP LTE C-V2X Phase 2 services. At the time this book was written, corresponding discussions were in an early phase and 3GPP RAN had approved two new V2X items in its 75th meeting (March 2017). The related LTE V2X R15 work scope is illustrated in Figure 5.22.

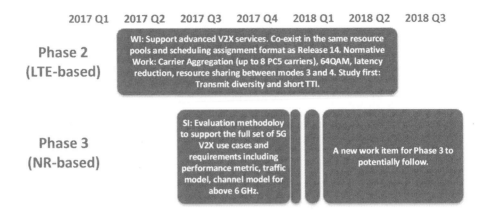

Figure 5.22: 3GPP V2X evolution

To summarize, at least the following features are expected to be defined in 3GPP Release 15:

- Sidelink carrier aggregation (up to 8 sidelink component carriers)
- Support of higher order modulation (64QAM: attempt to increase throughput per single CC)
- Pool sharing between eNB-controlled and UE-autonomous resource allocation modes (for example, Mode 3 and Mode 4), as illustrated below.

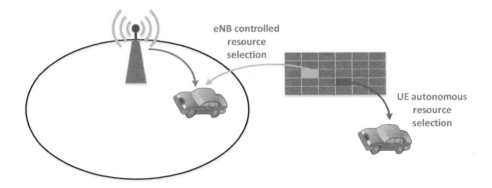

Figure 5.23: Resource selection mechanisms

- Latency reduction (Reduction of resource selection time)
- Potential design enhancements: Under feasibility/performance analysis study
- Support of sidelink short TTI
- Support of transmit diversity schemes (SC-FDMA based)

For 3GPP LTE V2X Phase 3, SA1 defined 25 eV2X use cases and categorized them into four main groups (TR 22.886). An overview is illustrated in Table 5.2. Advanced driving and extended sensor sharing are likely to be given higher priority and focus.

Table 5.2: Categorized V2X use cases

Use Case	Illustration	Description
Platooning		Vehicles dynamically form a platoon traveling together. Vehicles in the platoon obtain information from the leading vehicle to manage this platoon.
Advanced Driving		Vehicle/RSU shares its own perception data obtained from its local sensors with vehicles in proximity, which allows vehicles to coordinate their trajectories.
Extended Sensor		Exchange of data gathered through local sensors or live-video images among vehicles, RSUs, pedestrians, and the V2X server.
Remote Driving		Enables a remote driver or a V2X application to operate a remote vehicle.

Figure 5.24 outlines the differences between 3GPP LTE V2V Rel. 14 and the future 5G V2X.

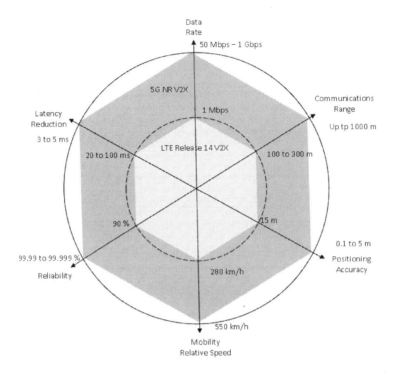

Figure 5.24: Comparison of LTE V2V R14 and 5G V2X

Related features under discussion include the following:
- Positioning is a key V2X service component
- 3GPP New Radio (NR) V2X: Impact on radio interfaces
- Platooning: Accurate vehicle positioning, URLLC sidelink (groupcast/broadcast), sidelink radio resource management, support of DL and UL communication
- Extended Sensors: High throughput sidelink, support of broadcast, multicast, unicast, URLLC sidelink
- Advanced Driving: Accurate vehicle positioning, URLLC sidelink, support of DL and UL communication, including URLLC
- Remote Driving: Ultra-reliable and low-latency DL/UL (Uu radio interface), high throughput UL is needed, no impact on sidelink air-interface
- 3GPP New Radio (NR) V2X Positioning: Main principles and benefits are illustrated in Figure 5.25.

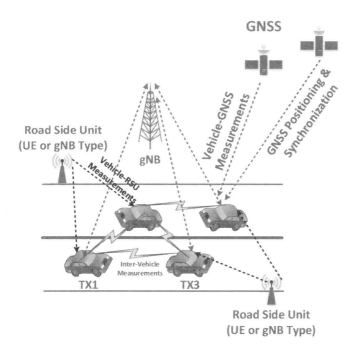

Figure 5.25: Principles for V2X positioning

– Role of NR V2X Positioning: Relative and absolute location with vehicle(s)/RSU(s)/gNB(s), broadcasting of location information, enabling cooperative location
– Technical Advantages: Wider bandwidths, angular information, improved accuracy (especially in urban), increased reliability/robustness (multiple location sources)
– Communication and positioning protocol: NR V2X technology components, dual band operation/communication (Figure 5.26), 5.9 GHz and 63 GHz

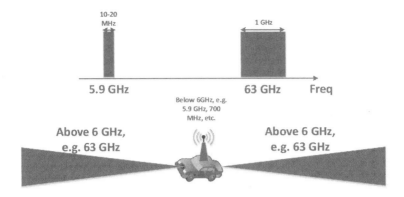

Figure 5.26: Example for dual operation of 5.9 GHz and 63 GHz ITS services

— Sidelink MIMO: Distributed antenna array systems with joint/distributed processing, radar and communication capabilities, advanced receivers (ISIC), receivers with advanced ISIC capabilities (friendly transmission format), multiple transmission and decoding attempts on V2V resources to improve reliability (see Figure 5.27).

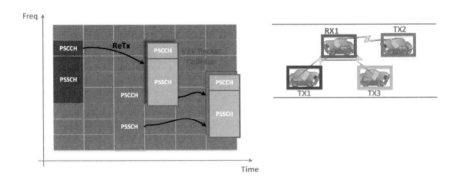

Figure 5.27: Multiple Transmission and Decoding Attempts to improve Reliability

— Sidelink radio-layer relaying: Multi-hop relaying/dissemination, sidelink ranging/positioning (see Figure 5.28).

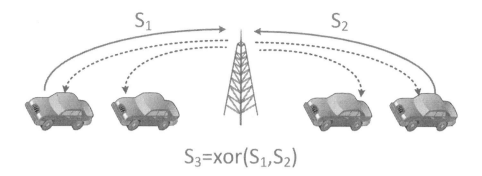

Figure 5.28: Information combining

- Estimation of time of flight and relative distance measurements, absolute coordinate refinement
- Centralized and distributed architecture: Intelligent SL resource selection, management and feedback mechanism, utilization of geo-location information
- Road Side Units: Reassessment of UE-type RSU role, mobile-edge computing
- Numerology: Increased subcarrier spacing, robust demodulation in high speed environment
- Congestion Control

References

Apratim Choudhury; Tomasz Maszczyk; Muhammad Tayyab Asif; Nikola Mitrovic; Chetan Math; Hong Li; Justin Dauwels (2016): 2016 IEEE 19th International Conference on Intelligent Transportation Systems (ITSC). Windsor Oceanico Hotel, Rio de Janeiro, Brazil, November 1-4, 2016. With assistance of Rosaldo Rossetti. Piscataway, NJ: IEEE. Available online at http://ieeexplore.ieee.org/servlet/opac?punumber=7784515.

IEEE International Conference on Communications; Institute of Electrical and Electronics Engineers; ICC; IEEE ICC (2016): 2016 IEEE International Conference on Communications (ICC). 2016 IEEE International Conference on Communications (ICC) took place 22-27 May, 2016 in Kuala Lumpur, Malaysia. Piscataway, NJ: IEEE. Available online at http://ieeexplore.ieee.org/servlet/opac?punumber=7502491

2014 IEEE 28th International Conference on Advanced Information Networking and Applications (AINA). Victoria, BC, Canada.

5GAA Automotive Association (2017): Comments of the 5G automotive association. Proposed Changes in the Commission's Rules Regarding Federal Motor Vehicle) Safety Standards; V2V Communications ET Docket No. NHTSA-2016-0126. 5GAA Automotive Association.

ERTICO; European Commission; 5G-PPP (2015): 5G Automotive Vision. Edited by ERTICO, European Commission, 5G-PPP.

The Innovation Imperative (2015): OECD Publishing.

TOYOTA InfoTechnology Center, U.S.A., Inc. (7/25/2016): Toyota ITC's Perspective on 5G.

Abboud, Khadige; Omar, Hassan Aboubakr; Zhuang, Weihua (2016): Interworking of DSRC and Cellular Network Technologies for V2X Communications. A Survey. In IEEE Trans. Veh. Technol. 65 (12), pp. 9457–9470. DOI: 10.1109/TVT.2016.2591558.

Bansal, Gaurav; Kenney, John B.; Weinfield, Aaron (2011): Cross-Validation of DSRC Radio Testbed and NS-2 Simulation Platform for Vehicular Safety Communications. In Gaurav Bansal, John B. Kenney, Aaron Weinfield (Eds.): Cross-Validation of DSRC Radio Testbed and NS-2 Simulation Platform for Vehicular Safety Communications. 2011 IEEE Vehicular Technology Conference (VTC Fall). San Francisco, CA, USA, pp. 1–5.

Bansal, Gaurav; Kenney, John B.; Weinfield, Aaron (Eds.) (2011): Cross-Validation of DSRC Radio Testbed and NS-2 Simulation Platform for Vehicular Safety Communications. 2011 IEEE Vehicular Technology Conference (VTC Fall). San Francisco, CA, USA.

Datta, Soumya Kanti; Haerri, Jerome; Bonnet, Christian; Ferreira Da Costa, Rui (2017): Vehicles as Connected Resources. Opportunities and Challenges for the Future. In IEEE Veh. Technol. Mag. 12 (2), pp. 26–35. DOI: 10.1109/MVT.2017.2670859.

Gallo, Laurent; Haerri, Jerome (2017): Unsupervised Long- Term Evolution Device-to-Device. A Case Study for Safety-Critical V2X Communications. In IEEE Veh. Technol. Mag. 12 (2), pp. 69–77. DOI: 10.1109/MVT.2017.2669346.

Hashem Eiza, Mahmoud; Ni, Qiang (2017): Driving with Sharks. Rethinking Connected Vehicles with Vehicle Cybersecurity. In IEEE Veh. Technol. Mag. 12 (2), pp. 45–51. DOI: 10.1109/MVT.2017.2669348.

Khoryaev, Alexey (4/5/2017): LTE and 6G NR V2X.

Matthias Pätzold (2017): Fifth-Generation Developments Are in Full Swing. In IEEE vehicular technology magazine, pp. 4–12. DOI: 10.1109/MVT.2017.2681978.

Protzmann, Robert; Schunemann, Bjorn; Radusch, Ilja: A Sensitive Metric for the Assessment of Vehicular Communication Applications. In : 2014 IEEE 28th International Conference on Advanced Information Networking and Applications (AINA). Victoria, BC, Canada, pp. 697–703.

Seo, Hanbyul; Lee, Ki-Dong; Yasukawa, Shinpei; Peng, Ying; Sartori, Philippe (2016): LTE evolution for vehicle-to-everything services. In IEEE Commun. Mag. 54 (6), pp. 22–28. DOI: 10.1109/MCOM.2016.7497762.

Sjoberg, Katrin; Andres, Peter; Buburuzan, Teodor; Brakemeier, Achim (2017): Cooperative Intelligent Transport Systems in Europe. Current Deployment Status and Outlook. In IEEE Veh. Technol. Mag. 12 (2), pp. 89–97. DOI: 10.1109/MVT.2017.2670018.

Su, Zhou; Hui, Yilong; Yang, Qing (2017): The Next Generation Vehicular Networks. A Content-Centric Framework. In IEEE Wireless Commun. 24 (1), pp. 60–66. DOI: 10.1109/MWC.2017.1600195WC.

Uhlemann, Elisabeth (2017): The US and Europe Advances V2V Deployment [Connected Vehicles]. In IEEE Veh. Technol. Mag. 12 (2), pp. 18–22. DOI: 10.1109/MVT.2017.2680660.

Wei, Hung-Yu (9/24/2015): From D2D to V2X.

Zhang, Ke; Mao, Yuming; Leng, Supeng; He, Yejun; ZHANG, Yan (2017): Mobile-Edge Computing for Vehicular Networks. A Promising Network Paradigm with Predictive Off-Loading. In IEEE Veh. Technol. Mag. 12 (2), pp. 36–44. DOI: 10.1109/MVT.2017.2668838.

Chapter 6
Infotainment

The way drivers and passengers interact with their vehicles and vice versa has evolved for many years and will continue to do so. It all began with simple vehicle radio and audio systems adapted to meet the needs of drivers and passengers in the vehicle environment. Later, other services got added, such as Germany's vehicle driver's radio information system for receiving traffic announcements from radio stations in the mid-1970s. Telematics features were added to these entertainment and news applications with the emergence of global positioning systems (GPS) in the 1990s. The radio data system (RDS), which got later renamed traffic message channel (TMC), implemented through radio broadcasting, followed in 1998 and made it possible to transmit traffic data right into the vehicle. All these services and corresponding features were enclosed in the vehicle radio.

But GPS was the entry point into a completely new period for vehicle infotainment and telematics, since it required additional computing storage and communication performance for digital data. These data processing capabilities were needed for processing the GPS and other sensor data, like an odometer to estimate the vehicle's exact position. The human-machine-interface (HMI) to interact with the GPS based telematics system and the vehicle infotainment and control system had to advance as well. It took advantage of the readiness of better high-resolution color displays and rotary-push knobs for instance. And nowadays, HMI systems operate with the support of touch screens, voice or gesture control.

Until now, vehicle human-machine interface (HMI), control, telematics and navigation, entertainment and infotainment and communication functions were separately implemented in dedicated system units like the above-mentioned vehicle radio. This changed when wireless networking and connectivity became a crucial technology in vehicles. The ongoing penetration of smartphones leaves little room for the vehicle ecosystem stakeholders to escape it. So, vehicle ecosystem stakeholders take on matters like the choice of constantly updated data, the option to integrate smartphones and tablets into vehicle HMIs and telematics, the possibility of internet access, or the opportunity to feed back vehicle data into the cloud. Initially the premium and higher end class of vehicles get HMI systems with interfaces for wireless networking and connectivity. Then this creates increasing demand from drivers and passengers for advanced user interface connectivity features in all vehicles classes. The smartphone and embedded networking and connectivity features evolve and enable finally a seamless integration of wireless networking and connectivity which pushes vehicles into an open V2X ecosystem.

The challenge for vehicle HMI ecosystem stakeholders is it to advance the HMI in a way that it stays easy and simple to use, flexible, functional and engaging for drivers and passengers, especially in competition with an ongoing complementary use of

DOI 10.1515/9781501507243-006

smartphones and tablets in the vehicle. For example, the HMI shall enable a seamless and disruption-free transition between cellular, Wi-Fi or satellite access links between in-vehicle and external communication or an aggregation of drivers' and passengers' smartphone links for fastest data download and upload for infotainment, telematics and control when available and needed.

There is an evolution in infotainment, telematics and vehicle control and diagnostics themselves. And there is the option to run all of them out of one single unit including C-LTE or DSRC networking and connectivity. Finally, this becomes a starting point for a common vehicle HMI evolution around the IVI toward connected infotainment, telematics and vehicle diagnostics, integrating all these current technology silos. Putting the domain's cockpit, safety, body and powertrain on a single server processing platform and running them on virtualized machines provides the flexibility and scalability requested by the vehicle ecosystem stakeholders.

In a vehicle, every window and seat motor, headlamp and trunk release has its own MCU which adds up to one hundred MCUs and sensors or actors interconnected via networking and connectivity in the vehicle. This topic gets even more challenging for automotive vehicle manufacturers now with the vehicle's connectivity to the internet and advanced safety features like lane or parking-assist and in-vehicle infotainment (IVI). The vehicle IVI system must offer additional features such as 360-degree video, back-up cameras and smartphone integration with the integration of supplementary units. So, a telematics control unit (TCU), a navigation control unit (NCU), a vehicle control unit (VCU) and a communication control unit (CCU) get implemented.

6.1 Infotainment

In-vehicle infotainment (IVI) is a broadly used vehicle industry term that refers to the vehicle system components providing radio and audio entertainment and news to drivers and passengers. IVI solutions advanced in vehicles with the introduction of navigation systems and global positioning systems in the 1990s. Then these systems progressed further with the climate control system, the software-defined intelligent instrument cluster, and the high-resolution touch displays with access to apps, voice control, and vehicle sensor and actor status to an in-vehicle experience (IVE) today. Connected in-vehicle infotainment systems (IVI) are a combination of entertainment and data delivery to vehicle drivers and passengers using multi-modal interfaces and connectivity. IVI functions include for example GPS navigation, video players, music streaming, short messaging, hands-free calling, in-vehicle internet and USB, Bluetooth, and Wi-Fi connectivity.

A current in-vehicle infotainment system is an integrated set of input and output (I/O) devices which are networked and communicate with each other using bus systems, such as CAN, and provide an interface between the driver, the passengers and the vehicle. I/O devices are a key hardware component of in-vehicle infotainment and offer a large number of physical human machine interfaces. Compared with the faceplate of past vehicle head units, the IVI faceplate starts to implement displays that come with touchscreens, force feedback and gesture recognition. Additionally, the driver and passengers are able to access data on connected devices such as MP3 players, smartphones, USB devices, and flash drives.

IVI implementations do not necessarily open wireless networking and connectivity capabilities as such, but make available interfaces for example smartphone integration, at least. This way IVI systems get wireless networking and connectivity functions to offer drivers and passengers many mobile services. Such IVI systems make use of flexible computing and communication platforms with processors running embedded operating systems. There is the choice to integrate a variety of radio access technologies like cellular wireless, Bluetooth, near field communication (NFC) and Wi-Fi. Software provides the scalability and flexibility to develop and add easily familiar personal computer and smartphone based applications and get access to a wide variety of development tools to develop and implement innovative IVI applications.

IVI system interfaces are knobs, buttons, keyboards, wheels, touch displays and gesture input today to control audio, video, networking and connectivity, telematics and vehicle functions. The vehicle cockpit instrument unit for IVI integrates all these into one unit. For example, a single touch display enables the control of vehicle's heating, ventilation and air conditioning (HVAC), vehicle interior light, seats, rear view and other camera views and infotainment. The IVI interfaces evolve currently with 3D and augmented reality (AR) navigation, multimedia support, additional device integration, high-speed connectivity, intuitive and multi-modal user interfaces (UI) and vehicle cloud services. This development of in-vehicle infotainment systems is driven by the need for safe, simple, efficient, and comfortable interfaces.

Whereas the simplicity of vehicle HMIs has been common for many decades, the current developments in the area of in-vehicle infotainment make them more sophisticated. Comparing past and present developments, the main differences are the aspects of data processing, storage and networking and connectivity. Human-machine interfaces get graphical user interfaces, complex control units and more and more speech- and gesture- recognition systems due to safety needs. The implementation of an increasing number of hardware and software components increases the complexity of IVI systems and usability has become a key design metric.

Today's vehicle manufacturers rely on smart in-vehicle infotainment systems to differentiate themselves. Vehicle drivers and passengers expect the same compelling user experience in their vehicles as they get when they use their tablets, smartphones and other computing devices. The overwhelming capabilities of computing and communication technologies to deliver an intuitive, appealing and fully connected user experience can be implemented very differently between vehicle manufacturers.

For instance, Audi's MMI use a 6.5-in. display with 400 × 240 pixels and supports features such as smartphone connectivity, address book, CD player, and TP memory.

In addition to that, the MMI Radio plus is equipped with two SD card readers, Bluetooth phone, a speech dialog system, and the ability to play MP3 files. MMI Navigation extends the variant MMI Radio plus with a DVD-based navigation system. TMC traffic information, and a speech dialog system allow controlling the address book, the phone and entering navigation destinations via voice commands. The variant MMI Navigation plus, which is standard in the A8 and A6 Avant, provides an 8-inch display with 800 × 480 pixels, a hard-disk-based navigation system with 3D map and Google Earth satellite views, a DVD drive, USB port, iPod interface, advanced driver assistance systems, and a speech dialog system with full-word input to control the navigation system, the phone, and the address book.

The Audi infotainment systems extend the range of available mobile online services, which are called Audi Connect. These services include Google Search, Google Street View, traffic information, news, and a weather forecast. In the Audi A8 and A7, it is possible to connect up to eight mobile devices at the same time to a WLAN hotspot provided by the infotainment system. In addition to those MMI variants, Audi offers a complete display called "virtual cockpit". It is a fully digital instrument cluster focused on the driver. In the 12.3 inch TFT display, all functions of a standard instrument cluster and the middle MMI monitor are combined.

Another example is the Buick or GMC IntelliLink (Chevrolet MyLink) which is an in-vehicle infotainment system with voice-activated beck and call. It integrates a driver or passenger smartphone into the vehicles' IVI as an audio player, navigation device or computer. IntelliLink has got a 7-inch, color touchscreen and offers speech recognition for example to safely initiate phone calls. The navigation system can get updates via smartphone. There are interfaces for USB and Bluetooth as well. GM Cadillac's CUE IVI system uses touch-sensitive buttons and an 8.0inch touchscreen and runs similar to the IntelliLink system. Apple CarPlay and Android Auto are supported as well with Bluetooth and a 4G LTE Wi-Fi hotspot which allows passengers to connect up to seven mobile devices, smartphones, and tablets to the vehicle's wireless internet connection.

In-vehicle infotainment, together with their HMIs, change rapidly and continue to grow in complexity. Combined with more and more new technologies from other domains, in-vehicle infotainment systems increase the number of mobile applications and services and comfort in modern vehicles. The range of driver and passenger expectations is also widening, from driving assistance toward high-quality entertainment. HMIs have to support graphical user interfaces based on 3D graphics and manage the multimodal interfaces. Integrated vehicle solutions with reduced design complexity and robust platform, simplifying the integration process, and enabling customization are asked for.

With the rising concerns over safety, complexity and diversity of networking and connectivity functionality in vehicles, these issues are becoming more and more critical. An IVI system becomes an integrator of various input and output devices which get linked and communicate with each other to enable interaction between the driver or passengers and the vehicle. The separation of the IVI system and the ECU system is a must do.

Automobile manufacturers started to aggregate functions within a single device in order to reduce complexity. The complete system could then be accessed via one graphical user interface with a hierarchically structured menu. At that point, premium vehicle manufacturers like Audi, BMW, and Mercedes-Benz presented their first in-vehicle infotainment systems combining informative and entertaining functionalities. Until today, the main subjects are multimedia (e.g., radio, mp3, and television), vehicle information (e.g., trip length, temperature), navigation, and telecommunication.

Due to the increasing complexity of the HMI, the usability of the interfaces has become a very important quality factor. Since the 1980s, standards have been defined to develop user interfaces with high usability. Soon it became apparent that for the automotive domain special user interface standards had to be established because automotive HMI differ in major points from HMIs in other domains. One big difference is the focus on user attention. Whereas in many domains the main task of the user is to interact with the application, with automotive HMIs driving must remain the highest priority. When the functionalities of infotainment systems increase, the causes for driver distraction increase, too. In addition, the cognitive load for performing a task can grow immensely. It has become more and more important over the years to ensure safety when developing automotive HMIs. Another difference to other user interfaces is that the devices in cars are normally at fixed positions and the user can only interact with them within a limited radius.

Modern smartphones provide powerful hardware with high-definition touchscreens and sensory input- and output-like compass and GPS. Relying on permanent internet availability, manifold functions and applications are possible at low costs. Furthermore, new apps can be installed easily. This leads to increased vehicle driver and passenger expectations, which carry over to in-vehicle infotainment systems, because driver and passengers compare their in-vehicle infotainment systems with other devices of their everyday life. These trends are a paradox for the vehicle ecosystem stakeholders. Highly innovative infotainment systems are expected in vehicles. Computing and communication technologies like internet access, apps and app updates, as well as the seamless integration of mobile devices in the vehicle become a self-evident part of these systems.

The number of infotainment functions that perform better using in-vehicle systems than by consumer electronic devices decrease. Some of the major challenges with regard of the impact of this trend are in relationship with the differences in the development and the product life cycles for automotive products and computing and communication electronics. In order to bridge the life cycle gap between mobile devices and in-vehicle infotainment systems, automotive and mobile device manufacturers have to cooperate with other vehicle ecosystem stakeholders. Exchange formats and interfaces have to be defined and flexible software architectures must be developed. Extendable HMIs have to be described in a way that they suit different input and output technologies as well as operational concepts found in different vehicle models. In contrast to conventional telematics services, where the web service only provides machine to machine interfaces and no form of data presentation, these apps require new forms similar to web technologies for the realization of HMIs.

HMIs get permanently connected to mobile devices and therefore to the Internet, and make use of server- or cloud-based services: This connectivity fosters not only the ability to change software easily even after roll-out, but also to reduce the constantly growing hardware demands. For example, hybrid HMIs with mobile devices provide possibilities to extend the functional range of current infotainment systems by integrating new functions from external sources. These HMIs are dynamically adaptive, depending on the habits and demands of the drivers and the driving situation, and get personalized. There are different setups possible in which infotainment systems, external devices, and web services take over different roles. This requires technologies such as Mirror Link, Apple Car Play, or Android Auto, which enable remote operation of mobile phone applications.

6.2 Telematics and control

The telematics system supports the driver in navigation. Traffic data are visually aggregated on the navigation map and traffic message channel (TMC) data input gets added. Whereas the vehicle control or configuration functions, e.g., climate control, seat function, and in-vehicle lighting, enable the driver and passengers to configure the vehicle easily and comfortably. For instance, the climate control panel is used for setting air conditioning parameters, such as the fan speed and the temperature. Although the infotainment faceplate and the climate control panel could often be located next to each other, they are controlled by separate ECUs and developed in different departments at present. Other controls are for driver assistance functions, electronic stability control, door lock, or voice control.

Present in-vehicle infotainment systems use IVI levers to control infotainment functions or driver assistance functions such as adaptive cruise control (ACC). The levers as well as the buttons and scroll wheels on the steering wheel can be assigned to a single function, such as to accept incoming calls, activate voice recognition, or change the audio volume. They can also be used as multi-purpose controls for navigating in lists displayed in the cluster or head-up displays like in the digital cockpits. Voice control enables the driver or passengers to input commands in natural language without taking hands off the wheel and eyes off the road.

Nowadays, the driver can access functions like music selection, destination input for the navigation, or even climate control changes with the vehicle's embedded voice recognition system. But beyond the in-vehicle solution also smartphone functions which let the user interact with music, social media or phone contacts via voice control are common like Apple's Siri, Google's Google Now, or Microsoft's Cortana.

A recent addition to the list of input devices are camera controls. They are used for monitoring the driver and for gesture recognition. It is possible to combine this with other functions that would require a camera, such as video telephony. The camera is located on the dashboard, in the instrument cluster or in the back mirror.

The optical channel is still the predominant output device. Displays are used to provide the driver and passengers with data about the current vehicle system state. Today, the common locations of the major displays are the head unit and the instrument cluster. Conventional instrument clusters consist of electro-mechanical tachometers, speedometers, odometers, oil gauges, etc. These are often complemented by a display used for showing information such as the on-board computer or ACC warning. Trend leads to free programmable instrument clusters without the classical mechanical components.

For example, the digital cockpit of Audi or PSA also has a representation of the tachometer and speedometer and is reconfigurable in size. Depending on which data are important in a special situation, the driver can change the graphical representation. For example, the vehicle rounded elements will become smaller when the driver needs a bigger few of the navigation system. Head-up displays (HUD), which were used in primitive versions in airplanes, are a recent innovation in vehicles. To increase accuracy and usability when operating menus, force feedback has been widely used to help the driver get some kind of haptic feedback. Force feedback can also be used in the steering wheel to provide some driver assistance systems, such as the lane departure warning system (LDW). The availability of these different kinds of displays enables the creation of situation-dependent presentation sets or individually adaptable presentation forms the driver can choose from. For example, there might be a route guidance mode, an audio mode, and a night driving mode. It might even be possible to create your own personal presentation profiles.

Vehicle systems with V2X networking and connectivity support vehicle drivers to navigate smoothly with enhanced navigation technologies, including the global positioning system (GPS), the Russian global navigation satellite system (GLONASS), the Chinese BeiDou navigation satellite system and the European Galileo. They are application enablers to enhance experiences with applications such as UI, infotainment, navigation and advanced driver assistance systems (ADAS) and provide reliable software updates, diagnostics, connectivity management, media management, human-machine interface (HMI), and apps frameworks. Robust connectivity is based upon multiple vehicle connectivity technologies with multiple wireless and wired interfaces across cloud, vehicle and in-vehicle systems.

There are many efforts to offer telematics services based upon V2X networking and connectivity on a subscription basis by vehicle manufacturers like GM, Daimler, BMW, Qoros or Tesla and service providers like Verizon or Vodafone. Telematics services are part of an application bundle as vehicle built-in features, together with safety, security, location-based services, data transfer, vehicle diagnostics or software updates over the air. Currently cellular wireless or Wi-Fi are the radio access technologies for infotainment, telematics and business model use cases like emerging vehicle fleet services. At present the focus of vehicle manufacturers is on the delivery of internet access via embedded LTE modem and Wi-Fi in the vehicle. In particular Wi-Fi is efficient for firmware updates and data intensive applications in the vehicle. But this provision of Wi-Fi through LTE is technically complex and has got regulatory issues for example in Europe. Therefore, Wi-Fi provided by drivers' or passengers' smartphones are alternatively used. There is a clear lack of harmonized common wireless networking and connectivity solutions in place yet to get the vehicle connected with the cloud for envisaged future use cases similar to automated and autonomous driving and gathering, fusing and processing vehicle data in real-time.

6.3 In-vehicle connectivity and networking

In-vehicle systems are rapidly advancing in complexity and diversity since a multitude of sensors, actors and processors are used in different parts of the vehicle for various functions which need to be connected. This has led to in-vehicle networks growing in both size and complexity in an organic fashion with many complex sand-boxed heterogeneous systems in a single vehicle. The bandwidth issue has been brought into sharp focus through the introduction of infotainment and camera-based advanced driver assistance systems (ADAS) since they require significantly more bandwidth than traditional control applications. While a lot of the current focus about the connected vehicle is on external V2X networking connectivity, the quickly growing complexity of in-vehicle infotainment and brought-in devices is generating an urgent need for next-generation in-vehicle networking technologies and architectures.

In-vehicle networks vary according to the data traffic types from low bandwidth control, real-time control, safety data, and infotainment up to driver assist cameras. Low-bandwidth control data need low bandwidth and low quality of service (QoS) and are not safety critical like comfort subsystems (seat and mirror control). Real-time control data are low bandwidth as well, but with high real-time and quality-of-service (QoS) like suspension and braking systems, ABS and traction control. Safety data are for example, adaptive cruise control using a camera, radar and LIDAR sensors and ultrasound parking sensors. Infotainment data encompasses entertainment and driver and passenger data services including navigation, camera feeds, audio and visual entertainment and internet. Driver assistance cameras require high bandwidth and high QoS such as reversing cameras, bird's eye view cameras and rear cameras.

In-vehicle connectivity and networking is by far mainly done via wireline at present for reasons of latency and bandwidth, reliability and safety and security. Besides real-time, another holdup for networking and connectivity in the vehicle is bandwidth. With the improvements of Ethernet as a wireline networking and connectivity option in the vehicle, this changes as well. For example, wireline Ethernet was enhanced additionally with real-time elements for the transmission of real-time audio and video signals towards IEEE audio video bridging over Ethernet (AVB) to become an option in vehicles to transmit media data. Furthermore, time sensitive networking (TSN) enables a substantial reduction in the transmission time and offers now an option for driver assistance functions and safety-relevant applications. In-vehicle networking and connectivity as shown in Figure 6.1 is dominated by legacy communication technologies such as CAN, FlexRay, local interconnect network (LIN), media oriented systems transport (MOST) and low-voltage differential signalling (LVDS) and yet are challenged by Ethernet.

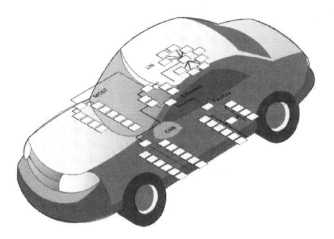

Figure 6.1: In vehicle networking and connectivity

The wireline networking and connectivity protocols of Table 6.1 support very different and specific vehicle function systems. LIN is used to provide a local interconnect network for lowest data rate functions to control mirrors, window lift, door switches and lock, HVAC motors, engine sensors, seat positioning motors and switches, wiper control, light and interface switches or vehicle interior lighting. LIN is a single-wire serial communications protocol supporting up to 12 network nodes and based on the SCI UART byte-word interface with 12 V supply voltage. The maximum data rate is 10 kbps. The controller area network CAN, for medium data rates, as a serial bus designed to provide a link between vehicle sensors and actuators. Up to 40 nodes can send a message at any time, when two nodes are accessing the bus together, arbitration decides who will continue. FlexRay is for high data rates and has time and event triggered behavior. It supports bus architectures as well as start network configurations. FlexRay provides error tolerance, speed and time-determinism performance needed for x-by-wire applications. Both, CAN and FlexRay are bus systems for safety critical function systems like the collision detection system or the brake-by-wire system. MOST is for very high data rates and is used, for example, for infotainment backbone to exchange multimedia data. And Ethernet gets applied for vehicle diagnosis and software upload.

Table 6.1: In-vehicle networking and connectivity protocols

Protocol	Bitrate in Mbps	Medium	Access control
LIN	Up to 0.020	Single wire	Polling
CAN	Up to 1	Twisted pair	CSMA/CA
FlexRay	Up to 10	Twisted pair/optical fibre	TDMA
MOST	Up to 24	Optical fibre	TDMA, CSMA/CA
Ethernet	Up to 1000	Twisted pair	CSMA/CD

The explosion of in-vehicle functionality such as multimedia systems, brake-by-wire, clutch-by-wire, steer-by-wire, drive-by-wire, throttle-by-wire, shift-by-wire transmission, internet access, GNSS, mobile communication, engine, body and power train intelligent control and data monitoring systems, increase the need for in-vehicle networking and connectivity. This function explosion and the related growth of the number of electronic control units (ECU) to more than 100 and of the number of software lines of code to more than 100 million prompt the vehicle manufacturers to look for new in-vehicle wired or wireless networking and connectivity technology options to achieve lower costs, reduced weight, simpler architectures, more reliability, higher bandwidth, lower latency, ease of use, functional safety and scalable and flexible repair and maintenance.

Connected vehicles have an unprecedented need for reliable and high-performance wireless connectivity. But in-vehicle infotainment (IVI), secure telematics, in-vehicle wireless gateways, and enhanced safety capabilities are very challenging. Nevertheless in-vehicle wireless networking and connectivity starts with infotainment use cases. One example are Bluetooth solutions which support for example remote key access, individual seat memorization, user-specific UI preferences and voice-activated functions. Bluetooth low energy (BLE) solutions enable always-on connections for example for tire pressure and security alerts. Another example is the connected rear seat infotainment, where rear-seat infotainment consoles are connected with the vehicle head unit to bring services such as video streaming and navigation to rear seat passengers. Here Wi-Fi IEEE 802.11 solutions provide multi-channel streaming support for multiple devices per vehicle.

Internet access in the vehicle and its use for general browsing and infotainment area prerequisite to the previous in-vehicle use case. A high quality data connection and good coverage make possible the reliable delivery of media and internet into a vehicle. This facilitates the provision of infotainment by internet as a standard feature of vehicles. A V2X networking and connectivity technology shall provide a consistent data rate that is high enough to support the chosen media. For internet browsing and general data there is the demand of at least 0.5Mbps data throughput, for high quality music streaming at least 1 Mbps, for standard quality video streaming at least 5 Mbps and for high quality video streaming at least 15 Mbps. Low latency is not very critical for media streaming, however, a latency of less than 100 milliseconds is needed for vehicle velocities up to 200 kilometers per hour. More challenging use cases are the integration of V2X networking and connectivity technology with in-vehicle connectivity as of the electronic stability control (ESC) linked with a data source providing data about slippery and icy roads or the use of micro, hyper local weather and precipitation data for windscreen wiper speed control.

The penetration of wireless networking and connectivity in vehicles is expected to grow rapidly, achieving full connectivity for all vehicles in 2025. Taking into account that almost every vehicle driver or passenger owns a smartphone with related apps we have already a 100 % penetration of V2X networking and connectivity technology today. Besides that, smartphones and apps have created an expectation among vehicle drivers and passengers that they get linked to the Internet 24/7, even in a vehicle driving around. The wireless networking and connectivity implementation options are embedded connectivity, tethered solutions and smartphone integration as summarized in Table 6.2.

Table 6.2: V2X options using smartphones

System component	Embedded	Tethered	Integrated
Wireless modem	In-vehicle built-in	Smartphone	Smartphone
UICC	In-vehicle built-in	Smartphone	Smartphone
Apps	IVI built-in	Smartphone	Smartphone
User interface	IVI HMI built-in	IVI HMI built-in	IVI HMI built-in or Smartphone HMI

In the embedded option, the wireless networking and connectivity module and the UICC are built into the vehicle, allowing connection to a cellular network without external devices. A typical example of an embedded option is a navigation system with real-time traffic. A tethered option makes use of a mobile smartphone or tablet for linking with an in-vehicle's infotainment system. Music streaming using Bluetooth is a typical case of this type of implementation. And there is finally the integrated option of connectivity. An example of this option is a software App integrated in the in-vehicle infotainment system and an on-board mobile smartphone or tablet. In this way, mobile smartphone or tablet functionality is accessed via a vehicle data processing platform including HMI control as well as access to the infotainment display and vice versa.

For example, Baidu CarLife, MirrorLink or Google's Android Auto enable the tethered option with Bluetooth DUN/PAN or Wi-Fi where the smartphone access is controlled via the vehicle HMI. Apple's CarPlay, connects the smartphone as well to the in-vehicle infotainment system and extends it by sensors to scan the vehicle surrounding to update the infotainment system accordingly and the smartphone to reflect the environment status. All these are examples of widely used integrated options nowadays.

For instance, Toyota Entune uses the drivers' smartphone for connectivity and extends the functionality of the smartphone toward the centrally mounted touch screen which is augmented by capacitive touch knobs and buttons. Entune doesn't provide control over basic vehicle functions, focusing instead on integrating smartphone apps into the driving experience. Entune is comprised of CD, AM/FM HD Radio, SiriusXM satellite radio, a USB port, an auxiliary input, and Bluetooth audio. Bluetooth is used by Entune to access streaming music from iHeartRadio, Pandora, and Slacker via a gateway Entune app on an iOS or Android smartphone.

6.4 Conclusions

Vehicles with wireless networking and connectivity offer more than just infotainment, they provide richer multimedia features and in-vehicle internet access, to enable innovative telematics and navigation, support ADAS, safety and predictive maintenance capabilities. In-vehicle infotainment, telematics and control applications already make use of ubiquitous wireless networking and connectivity and therefore play their role altering vehicles into mobile services that provide them, their drivers and passengers and the whole ecosystem, opportunities to disrupt role models. The penetration rate of wireless networking and connectivity in vehicles increases steadily, and the kind of their integration are going to drive or delay how soon vehicles become a key part of the Internet of Things and change their role model in the ecosystem.

In-vehicle infotainment as a blend of data provided by telematics and in-vehicle entertainment systems, is a key accelerator of autonomous and automated driving vehicles. Obviously, one of the great attractions of autonomous and automated driving vehicles will be the opportunity to do something other than driving the vehicle, whether that means text messaging, watching a video, or talking on the smartphone. The behavior of the current in-vehicle infotainment and HMI systems is statically predefined.

In the future, there is a function prerequisite to personalize and adapt the in-vehicle infotainment, telematics and control system. The drivers and passengers are able to download, install, and update software for their infotainment, telematics and control system, ranging from simple stand-alone apps to new design styles that adjust the look and feel of the in-vehicle environment. This function involves wireless networking and connectivity and is a must have for example to adapt automated and autonomous vehicles which get asked for by passengers out of vehicle fleets and get adapted with their personal profiles. With this function to install, update or integrate third-party applications, the vehicle provider has to ensure that these apps are presented and get integrated in an appropriate manner within the existing storage, computing and input and output devices in the vehicle.

With these variably equipped in-vehicle infotainment, telematics and control, the quantity and functional range of available input and output devices differ from vehicle to vehicle, thus leading to an increased need for flexibly designed HMIs that are able to adapt to the respective context of use cases and scenarios. The objective is to create HMI concepts not only for one type of vehicle, but for a whole brand or model range of vehicles the driver may use. This also includes corresponding concepts for smartphone apps, web pages, or portals that belong to the vehicle sharing solution. One example is an ADAS function with the prioritization of warning messages and the definition of when and how these warnings are presented to the driver. With the increased wireless networking and connectivity of different vehicle components in vehicles and improved sensor systems, it is also possible to add more dynamically adaptive HMI based on knowledge of the current driving situation or the current driver.

The horizon of in-vehicle functions and applications gets broaden and many new services like remote diagnostics, payment services, or infotainment and entertainment services become viable with wireless networking and connectivity. Integrating the latest infotainment, telematics and control functions and presenting them in an appealing manner in complex HMIs leads to an increased demand for processor performance and memory capacity while cost pressure remains intense. At the same time, new and improved forms of interaction require incorporating new hardware elements such as approximation sensors, control elements including display capabilities, multi-touch displays, or touchpads providing haptic feedback.

Today's infotainment systems are rather closed systems that come with a statically defined set of vehicle manufacturer defined functions. Their interaction with external devices is limited at the moment to selected functionalities or even selected connectivity. Therefore, in future HMI and IVI systems, you can expect better connectivity, which will enable new functionality and customer value with less borders.

There are individually configurable digital instrument clusters which are first steps toward a higher degree of infotainment personalization. Downloadable content data and functionalities will enable drivers and passengers to individually extend the functional range of their HMI systems in the future. And the human–machine interaction in a vehicle is pushed to a next level by successfully combining natural language speech commands and touchpad gestures with coherent audio-visual feedback of the system.

The separation of short-lived hardware components, such as a graphics processing unit, main processor, or memory from long-lived parts, such as audio amplifier or CAN transceiver are part of variable hardware architectures as vehicle manufacturers want the opportunity to upgrade their existing infotainment system by not exchanging the boards. And industry standards like AUTOSAR and GENIVI allow development costs to be kept under control for both vehicle manufacturers and suppliers, as there is no further need to develop new adaptation layers in each new partnership. At the same time, they make it easier for an OEM to change to another supplier.

Furthermore, these apps have to comply with certain standards assuring minimal driver distraction, and they have to be seamlessly included into the dynamic HMI adaptation process. Some in-vehicle infotainment systems may violate the distracted driving laws of many countries. So, in addition to the necessary technical solutions, clear legal directives need to be established to define exactly which function of an in-vehicle infotainment, telematics or control system is made available to which vehicle driver and passengers under which specific driving conditions.

To accelerate the deployment of autonomous passenger vehicles, countries' distracted driving laws need to be reviewed and brought globally into regulation that appropriately permits telematics and infotainment systems. In China, Europe as well as the United States, infotainment and ADAS applications require changes to existing laws and greater regularity of laws. In addition, the challenge of safety validation rows further against the background of the increasing number of functions and widening ranges of variants and versions for each vehicle model, which reaches easily more than 150 variants with components for infotainment, telematics and control.

There are some distinctive applications that are hard to replace with external applications and devices. Driver assistance applications and all sorts of in-vehicle settings and comfort applications are major examples. Looking into the future, driver assistance applications may not be needed anymore for automated and autonomous driving vehicles. Access to infotainment and the possibility to work or relax or just be connected while driving is likely to be the major motivation for an individual to go for highly automated driving. This likely results in a major change in their lifestyle and improvement of their quality of life. It could also make long-distance commuting by vehicle more attractive, and thus offer a wider choice of residency location. Regulatory change would be needed in some countries to allow the use of infotainment while driving. The interplay between levels of automation and engagement in non-driving-related activities is a specific issue.

Integrated in-vehicle networking and connectivity solutions like Apple CarPlay, Android Auto or Baidu CarLife are thought to provide an efficient and compelling connected user experience for drivers and passengers. But the integration with the vehicle manufacturers' embedded infotainment system is a trade-off and does not lead to a harmonized and simple overall user experience of the infotainment system. The integration has to be set up and launched and increases the IVI system complexity creating often issues around IVI feature duplication, voice recognition and inconsistent HMI behavior.

More and more vehicles use the availability of at least one smartphone in the vehicle to provide connectivity and networking on top of integrated wireless connectivity and networking. There are Wi-Fi access points integrated servicing the passengers and driver in the vehicle connected to cellular networks via vehicle integrated wireless modems or driver or passenger smartphones. The use of smartphones, tablets and wearable form factors such as eyewear and smart watches requires to re-think the in-vehicle networking and connectivity standards.

References

AUDI AG (11/14/2013): Audi connect manual.

Car Connectivity Consortium (7/19/2017): Building Car Data Ecosystem.

Nuance (6/16/2017): Infographic What do passengers what.

Sasken (2/1/2017): Digital clusters: Virtualization with in-vehicle infotainment.

Berkner, Frank; Weller, Henri W.; Bauerschmidt, Werner (4/15/2015): Auto Infotainment - Driving the Course for Connectivity.

Bharati, Sailesh (2017): Link-layer cooperative communication in vehicular networks. New York NY: Springer Science+Business Media.

Hodgson, James; Bonte, Dominique (2016): HMI in autonomous driving. ABIresearch.

Jones, Matt (7/13/2016): Connected cars need connectivity and more.

Lanctot, Roger (2016): Wi-Fi and Verizon Transforming Telematics. Strategy Analytics.

Lanctot, Roger (2017): TCU Battleground. Strategy Analytics.

Maurer, Markus; Gerdes, J. Christian; Lenz, Barbara; Winner, Hermann (2016): Autonomous driving. New York NY: Springer Berlin Heidelberg.

Meixner, Gerrit; Müller, Christian (2017): Automotive user interfaces. New York NY: Springer Berlin Heidelberg.

Mourad, Alaa; Muhammad, Siraj; Al Kalaa, Mohamad Omar; Refai, Hazem H.; Hoeher, Peter Adam (2017): On the performance of WLAN and Bluetooth for in-car infotainment systems. In Vehicular Communications. DOI: 10.1016/j.vehcom.2017.08.001.

Olariu, Stephan; Weigle, Michele Aylene Clark (2009): Vehicular networks. From theory to practice / edited by Stephan Olariu, Michele C. Weigle. Boca Raton: CRC Press (Chapman & Hall/CRC computer and information science series).

Schramm, Dieter; Bardini, Roberto; Hiller, Manfred (2014): Vehicle Dynamics. Modeling and Simulation. Berlin, Heidelberg: Springer Berlin Heidelberg; Imprint: Springer.

Schreiner, Chris (2016): Early Apple CarPlay and Android Auto Integrations Bring Mixed UX Results. Strategy Analytics.

Wang, Yunpeng; Tian, Daxin; Sheng, Zhengguo; Wang, Jian (2017): Connected vehicles system. Communications, data, and control. Boca Raton: CRC Press.

Watzenig, Daniel; Horn, Martin (2017): Automated Driving. Safer and More Efficient Future Driving. Cham: Springer International Publishing; Imprint: Springer.

Winner, Hermann; Hakuli, Stephan; Lotz, Felix; Singer, Christina (Eds.) (2016): Handbook of Driver Assistance Systems. Basic Information, Components and Systems for Active Safety and Comfort. Cham: Springer International Publishing.

Chapter 7
Software Reconfiguration

This chapter discusses the question: Why should we employ software reconfigurability for vehicle communications?[1] Looking at software reconfigurability for radio systems in general, there is a clear trend in the telecommunication industry to evolve from classical, static radio systems toward reconfigurable radio systems (RRS). There are several reasons for this trend. First, the lack of spectral resources forces manufacturers to exploit novel technological trends in order to meet 5G-related promises in terms of quality of service, latency, reliability, and so on. Software reconfigurability enables the provision of corresponding feature updates. Second, the fast evolution and heterogeneous nature of the radio environment, combined with an ever-increasing computation power in mobile devices calls for new ways of ensuring that the diverse environment is exploited in the best possible way. Software reconfigurability is the key to dynamically adapting any target device to the specific location-dependent needs of its owner through the installation of tailored and targeted software components.

While these generic benefits are also appealing to the vehicular communications context, the basic challenge in this space relates to the longevity of vehicles and their radio components. The wireless industry is currently used to cycles of approximately two years for consumer products. Smartphones, for example, are typically replaced after this period of time. Vehicles, on the other hand, are designed for a lifetime of at least ten years or more. In some cases, vehicles are maintained in good condition over twenty, thirty, or even more years. This context leads to two major challenges.

First, the vehicular communication standards (such as 3GPP LTE C-V2X) are most likely evolving substantially over the lifetime of a vehicle. It is therefore important to have the technical possibility to update components of the built-in communications platform. It may not be possible to fully update the platform to comply with all latest generation features, but the most essential, safety-relevant components definitely need to be available. A software update over-the-air (OTA) avoids a vehicle recall to a service center and reduces the cost and time required to implement the new features.

Second, vulnerabilities and bugs of the original vehicular communications system are likely to be discovered and exploited over a ten-year time span. Such vulnerabilities may be inherent to the communication standard itself or due to specific

---

[1] Contributors: Vladimir Ivanov (State University of Aerospace Instrumentation, Russia), Seungwon Choi (Hanyang University, Seoul, Korea), Young-seo Park (Samsung Electro-Mechanics, Suwon, Korea), Yong Jin (Hanyang University, Seoul, Korea), Kyunghoon Kim (Hanyang University, Seoul, Korea), Heungseop Ahn (Hanyang University, Seoul, Korea), Scott Cadzow (C3L (Cadzow Communications Consulting Ltd), Sawbridgeworth, UK), Francois Ambrosini (IBIT Ambrosini UG (Haftungsbeschränkt), Munich, Germany)

implementation choices of a component manufacturer. The damage caused by a malicious exploitation of such a vulnerability can possibly threaten the safety and even the lives of the vehicle and its passengers. It is therefore essential to immediately correct such vulnerabilities when they are detected. Software upgrades are a useful tool to address and resolve such vulnerabilities at short notice. In the sequel, we will discover how the ETSI Software Reconfiguration framework defined by ETSI EN 302 969, EN 303 095 and EN 303 146-1/2/3/4 (see references) is exploited to provide exactly the software reconfiguration features identified above.

7.1 Issues to Be Addressed by Software Reconfiguration

We have addressed the great need for software reconfiguration above. But which are the specific, technical challenges that are solved by the ETSI software reconfiguration framework defined by ETSI EN 302 969, EN 303 095 and EN 303 146-1/2/3/4? In the following sections, we elaborate on a number of specific technical use cases that are directly addressed by the ETSI framework.

7.1.1 Problem statement 1: How to transfer and install radio software components to a target platform in a secure way

The ETSI software reconfiguration solution introduces a multitude of features. While the overall solution supports implementations of extended capabilities, a sub-set of these features (as defined by ETSI EN 303 146-1) is sufficient to provide the possibility to load novel software components to a target platform, to install and execute, and to uninstall such components in a secure way (as defined by ETSI TR 103 087 and TS 103 436).

7.1.2 Problem statement 2: How to enable a user to access new software components

The ETSI software reconfiguration solution supports a so-called RadioApp Store, an entity that offers access to a selection of radio software components. A user is able to access this store to identify all available software components and to download and install any selected component. Only those software components that have been previously tested and validated and are included in the Declaration of Conformity (DoC) of the target platform are visible to the user. Beyond the individual download of the RadioApp Store, the ETSI approach also allows for a massive deployment by upgrading all concerned devices of a given type.

7.1.3 Problem statement 3: How to deal with device certification in the context of novel radio software components

The ETSI software reconfiguration solution allows for the installation of new software components that alter the radio behavior of a target platform. A continued operation is only possible if the modified platform has been tested and validated and the responsible party (i.e. the manufacturer) makes a Declaration of Conformity (DoC) available that comprises the combination of the hardware and the new software components.

7.1.4 Problem statement 4: How to achieve software portability and execution efficiency

The ETSI software reconfiguration solution addresses the problem of how to make software portable to a multitude of distinct target platforms, such as smartphones of different manufacturers, and so on. The ETSI software reconfiguration solution introduces an efficient abstraction method based on a radio virtual machine approach that first creates a generic representation of a radio algorithm that, in the next step, is optimized for the target platform. The ETSI approach thus inherently provides high execution efficiency by omitting a middleware (as employed by the software communications architecture, for example).

7.1.5 Problem statement 5: How to enable a gradual evolution toward software reconfigurability

Legacy software reconfiguration solutions typically assume that entire radio access technologies (RATs) are being loaded through software onto a target platform. The ETSI software reconfiguration solution does not require an entire application to be replaced. Rather, the ETSI solution allows for a gradual replacement or re-parameterization of selected (hardwired) components. The particular components being available for replacement by software components are chosen by a manufacturer, and this selection can be modified over time. The manufacturer is able to manage the level of reconfigurability of the platform in a gradual and controlled way.

7.2 The Regulation Framework: Relationship Between Software Reconfigurability and the Radio Equipment Directive

In the European context, any radio device must comply with the relevant regulation directive. From 1999 to 2017, the European Radio Equipment and Telecommunications

Terminal Equipment Directive (R&TTE Directive) was in force. The R&TTE Directive did not explicitly exclude software reconfigurability of radio parameters, but no supporting articles were introduced either. This situation has changed with the publication of the new Radio Equipment Directive (RED) in which Articles 3(3) (i) and 4 are specifically designed to support software reconfiguration technology as shown in Table 7.1.

Table 7.1: Radio Equipment Directive, Articles 3(3) (i) and 4 on Software Reconfiguration of Radio Parameters

---

Article 3: Essential requirements

...

3. Radio equipment within certain categories or classes shall be so constructed that it complies with the following essential requirements:

...

(i) Radio equipment supports certain features in order to ensure that software can only be loaded into the radio equipment where the compliance of the combination of the radio equipment and software has been demonstrated.

Article 4: Provision of information on the compliance of combinations of radio equipment and software

Manufacturers of radio equipment and of software allowing radio equipment to be used as intended shall provide the Member States and the Commission with information on the compliance of intended combinations of radio equipment and software with the essential requirements set out in Article 3. Such information shall result from a conformity assessment carried out in accordance with Article 17, and shall be given in the form of a statement of compliance which includes the elements set out in Annex VI. Depending on the specific combinations of radio equipment and software, the information shall precisely identify the radio equipment and the software which have been assessed, and it shall be continuously updated.

Articles 3(3) (i) and 4 indeed clarify the regulation with a specific focus on how device certification can be achieved in a horizontal market; hardware and software (which alters radio characteristics of a device) manufacturers being independent entities. These articles require that one party needs to take full responsibility of the combination of hardware and software. In practice, this is likely going to be the hardware manufacturer or device importer. This entity is finally signing the Declaration of Conformity of a device and is thus taking full responsibility in case of any malfunction.

This regulation solution leads to a subsequent question: How should the certification of a device be handled in the case of new software components being developed after the market introduction (post-sale) of the original hardware? Strictly speaking, a new Declaration of Conformity needs to be issued when a new software component (which alters radio characteristics of a device) is made available to the user. The RED offers a solution in the context of Annex VII as shown in Table 7.2.

Table 7.2: Radio Equipment Directive, Annex VII

ANNEX VII
SIMPLIFIED EU DECLARATION OF CONFORMITY

The simplified EU Declaration of Conformity referred to in Article 10(9) shall be provided as follows: Hereby, [Name of manufacturer] declares that the radio equipment type [designation of type of radio equipment] is in compliance with Directive 2014/53/EU.

Using Annex VII of RED, the typical process for introducing and maintaining software reconfigurable equipment to the market is as follows: First, the entity responsible for the combination of hardware and software (typically, the hardware manufacturer or importer) issues an initial Declaration of Conformity and introduces the equipment to the market. Following Annex VII of RED, the equipment contains a pointer to a web address where the Declaration of Conformity is available. Second, an independent software-developing entity creates a new software application that affects the radio characteristics of the equipment. It is sent to the overall responsible entity for further processing—it is not yet available to the user. Third, the overall responsible entity checks the software component. In case it is decided to make it available to users, the overall responsible entity signs and issues a new Declaration of Conformity that specifically includes the operation of the new software component. This new Declaration of Conformity is made available at the web address indicated in the equipment in accordance with Annex VII of RED. Then, the user may download, install, and execute the application.

The RED software reconfiguration framework thus perfectly addresses the requirements of modifying radio characteristics of equipment through software. It is important to understand that RED only addresses radio behavior of an equipment and the corresponding essential requirements—non-radio related reconfiguration does not require any specific regulation provision, so there is no change to any previously available framework. Also, at the time when this book was written, Articles 3(3) (i) and 4 were not yet in force. The European Commission is in the process of setting up an expert group which is expected to work toward so-called implementing acts and delegated acts that will finally activate Articles 3(3) (i) and 4.

7.3 Mobile Device Reconfiguration Classes: How the Level Of Reconfigurability Will Grow Over Time

Software reconfiguration represents a new paradigm in radio equipment design. It will be some time until a fully flexible, highly efficient platform will finally be commercially available. Rather, it is expected that a gradual increase in flexibility will be applied. For this purpose, ETSI has defined mobile device reconfiguration classes (MDRCs) in ETSI EN 302 969 as illustrated in Table 7.3. The objective is to have a clear definition of the capabilities of a specific platform in order to address technical, certification, and security issues. These may indeed differ between the various MDRCs. While the exact definitions of MDRCs are given in ETSI EN 302 969, examples are used below in order to facilitate basic understanding.

Table 7.3: Mobile device configuration classes

Reconfiguration type	Mobile Device Reconfiguration Class	
No reconfiguration	MDRC-0	
No resource share (fixed hardware)	MDRC-1	
Pre-defined static resources	MDRC-2	MDRC-5
Static resource requirements	MDRC-3	MDRC-6
Dynamic resource requirements	MDRC-4	MDRC-7
	Platform-specific executable code	Platform-independent source code or IR

MDRC-0 (no reconfiguration) and MDRC-1 (no resource share—fixed hardware) represent today's commercial equipment. MDRC-0 does not support any reconfiguration at all and thus corresponds, for example, to a legacy Wi-Fi modem that cannot be switched to any other RAT. MDRC-1 still relies on fixed hardware implementations (ASIC types of chip designs, usage of static software); however, this reconfiguration class allows the switching between multiple distinct RATs and/or the simultaneous operation of a multitude of RATs.

MDRC-2 to MDRC-7 represent classes that enable software radio reconfiguration. Two columns are introduced in order to differentiate between two types of code: either platform-specific executable code (to be used on the target platform as-is), or platform-independent source code or intermediate representation (IR) code (which is further processed on the target platform through back-end compilation before execution).

In the pre-defined static resources cases (MDRC-2 and MDRC-5), any software component has a fixed allocation to specific computational resources. For example, a specific DSP among multiple DSPs is pre-defined for the code execution during compilation time. This approach is advantageous from a certification and testing perspective, since the final configuration is identical every time the equipment is used.

For static resource requirements (MDRC-3 and MDRC-6), the requirements are defined in a fixed way during design time. The need for a dedicated DSP for a piece of code may be identified. However, the specific DSP to be selected for the code execution is only identified during the installation of the corresponding software component and may differ each time the equipment is used.

In the final stage, called dynamic resource requirements (MDRC-4 and MDRC-7), any software component is dynamically mapped to any available computational resource during run-time. This approach typically leads to the highest level of efficiency, but also implies a highly unpredictable configuration of the equipment.

Note that the ETSI software reconfiguration method allows a gradual, step-wise approach to software reconfiguration. As a first step, for example, the manufacturer may choose to add spare computational resources (FPGA resources, etc.) to a hardwired (ASIC) implementation. Whenever required, some selected hardwired components could be replaced through software updates. In the future, a platform may employ more software-based components; consequently, it offers further post-sale reconfiguration capabilities. The definitions of MDRCs described above are summarized in Table 7.4.

Table 7.4: Summary of mobile device reconfiguration classes

Class	Multi-radio system	Resource share (Among radio applications)	Resource manager	Multi-tasking	Resource measurement	Resource allocation
MDRC-0	No	No	No	No	Design time	Design time
MDRC-1	Yes	No	No	No	Design time	Design time
MDRC-2	Yes	No*	Yes**	Yes***	Design time	Design time
MDRC-3	Yes	Yes	Yes	Yes	Design time	Run-time
MDRC-4	Yes	Yes	Yes	Yes	Design time	Run-time
MDRC-5	Yes	No*	Yes**	Yes***	Design/install time	Design/install time
MDRC-6	Yes	Yes	Yes	Yes	Design/install time	Run-time
MDRC-7	Yes	Yes	Yes	Yes	Design/install time	Run-time

*Resource share can exist among Radio Access Technologies (RATs) in a given radio application.

**This is for a fixed resource allocation only. Resource management and resource allocation among RATs (in a single RA) are pre-determined in a static manner by the radio application provider.

***Multi-tasking in this case is for multiple RATs within a single radio application.

7.4 The ETSI Reconfiguration Ecosystem

Figure 7.1 provides a high-level illustration of the reconfigurable mobile device (MD) architecture reference model for multi-radio applications as defined by ETSI EN 303 095. The reconfigurable MD architecture includes at least a Radio Computer for the processing of radio applications (RadioApps), software components that alter the radio characteristics of the target device. In the example in Figure 7.1, the dotted part belongs to either Radio Computer or Application Processor, depending on the specific implementation.

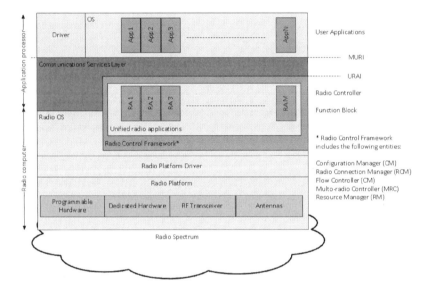

Figure 7.1: Reconfigurable mobile device (MD) architecture reference model for multi-radio applications

Following the illustration in Figure 7.1, the operation of the application processor is performed by a given operating system (OS). In the sequel, we use radio OS (ROS) for the operating system of the radio computer. The AP includes the two components driver, which has the purpose of activating the hardware devices (such as the camera, speaker, and so on) on a given mobile device (MD) and a non-real-time OS for the execution of non-real-time software reconfiguration functions. The Radio Computer includes two components: ROS as a real-time operating system, and a radio platform driver which is a hardware driver for the ROS to interact with the radio platform hardware.

The system architecture for a radio computer is illustrated in Figure 7.2 and Figure 7.3. Some of the entities included in the figures may be located in the cloud or any other external entity in order to off-load processing from the concerned mobile devices. To illustrate this offloading principle with an example, the back-end compiler in Figure 7.2 is moved into the cloud, as illustrated in Figure 7.3.

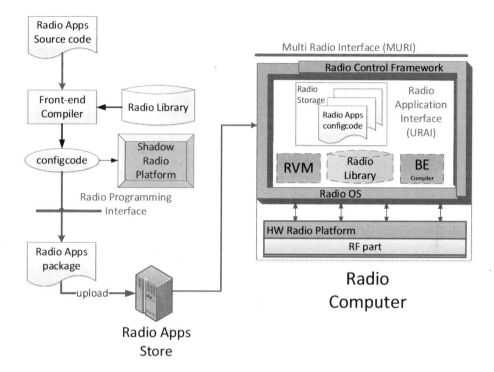

Figure 7.2: System architecture for radio computer where radio library and back end (BE) compiler are included within the radio computer

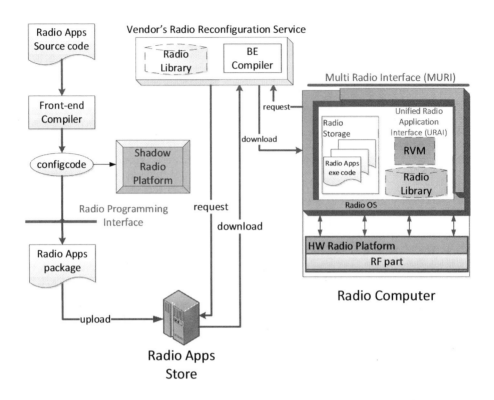

Figure 7.3: System architecture for radio computer where radio library and BE compiler are provided at a cloud outside the radio computer

Finally, Figure 7.4 illustrates the reconfigurable mobile device architecture and related interfaces enabling software reconfiguration.

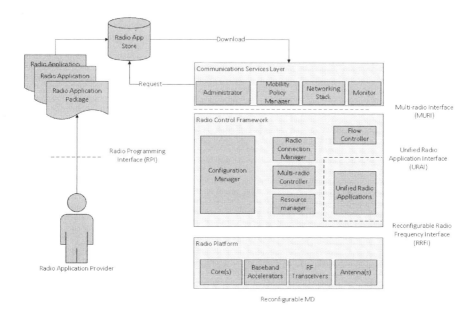

Figure 7.4: Standard reconfigurable mobile device architecture and related interfaces

As outlined in further detail in ETSI EN 303 095, a reconfigurable mobile device can execute the radio application (RA) code consisting of various functional blocks, for which the granularities might be all different depending upon hardware platform vendors. Depending on the features provided by mobile device manufacturers, the third-party software manufacturer develops the entire or partial RA code using the standard programming interfaces, as depicted in Figure 7.4. A modular software approach is applied in order to maximize the reusability of software components. The evolution of RATs can be supported by adding and/or replacing the functional blocks on a given hardware platform.

As shown in Figure 7.4, the following four components can be identified: a communication services layer (CSL) with the four logical entities—administration, mobility policy manager, networking stack, and monitor; the radio control framework (RCF) with five logical entities—configuration manager, radio connection manager, multi-radio controller, resource manager, and flow controller; and unified radio applications (URA) and radio platform consisting of RF transceiver, baseband, and so on.

These four components consist of software (CSL, RCF) and/or hardware (radio platform) entities and are interconnected through well-defined interfaces as defined in ETSI EN 303 146-1/2/3/4. The communication services layer (CSL) is a layer related to communication services that supports both generic and specific applications related to multi-radio applications. CSL includes the four entities—administrator, mobility policy manager (MPM), networking stack, and monitor.

As a minimum set of services, the administrator entity includes functions to request installation or uninstallation of URA, and creating or deleting instances of URA. This typically includes the provision of information about the URA, their status, etc. Furthermore, the administrator includes two sub-entities—the administrator security function (ASF), and the RRS configuration manager (RRS-CM). At a minimum, the MPM includes functions for monitoring the radio environments and MD capabilities, to request activation or deactivation of URA, and to provide information about the URA list. It also makes selections among different radio access technologies, and discovers peer communication equipment and the arrangement of associations. As a minimum set of services, the networking stack entity includes functions for sending and receiving user data. The monitor entity, at a minimum, includes functions to transfer information from URA to the user or to the proper destination entity in MD.

The radio control framework (RCF) provides a generic environment for the execution of URA, and a uniform way of accessing the functionality of the Radio Computer and individual RAs. RCF provides services to CSL via the multi-radio interface (MURI). The RCF includes the five entities for managing URA: configuration manager (CM), radio connection manager (RCM), flow controller (FC), multi-radio controller (MRC), and the resource manager (RM).

As a minimum set of services, the CM includes functions for installing and uninstalling and creating and deleting instances of URA, as well as management of and access to the radio parameters of the URA; the RCM includes functions for activating and deactivating URA according to user requests, and the management of user data flows, which can also be switched from one RA to another; the FC includes functions for sending and receiving user data packets and controlling the flow of signaling packets; the MRC includes functions to schedule the requests for radio resources issued by concurrently executing URA, and to detect and manage the interoperability problems among the concurrently executed URA; and the RM includes functions to manage the computational resources, to share them among simultaneously active URA, and to guarantee their real-time execution.

Note that the RCF, which represents functionalities provided by the Radio Computer, requires all RAs to be subject to a common reconfiguration, multi-radio execution, and resource sharing strategy framework (depending on the concerned MDRC). Since all RAs exhibit a common behavior from the reconfigurable MD perspective, those RAs are called URAs. The following services are accessible through the Unified Radio Application Interface (URAI), which is an interface between URA and RCF: Activation and deactivation, peer equipment discovery, and maintenance of communication over user data flows.

Figure 7.5 illustrates the integration of LTE C-V2X technology and the ETSI software reconfiguration framework. It is indeed straightforward to add existing vehicular system components to the CSL of the ETSI software reconfiguration technology. This platform provides all required components for addressing the software reconfigurations needed.

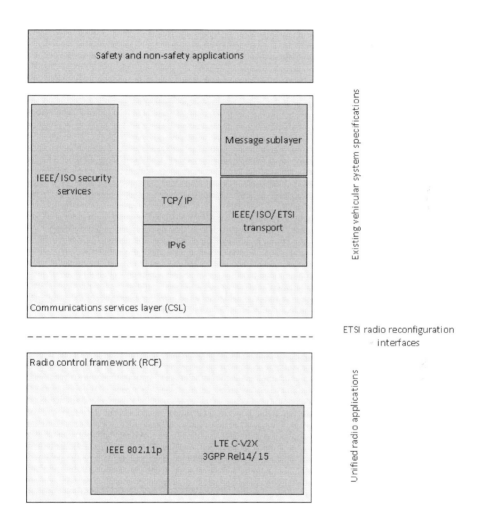

Figure 7.5: Architecture integrating 3GPP LTE C-V2X and the ETSI software reconfiguration framework

7.5 Code Efficiency and Portability Through the Radio Virtual Machine

The ETSI software reconfiguration solution is specifically designed for the requirements of commercial mass-market equipment. In order to achieve high efficiency in terms of power consumption and computational complexity, ETSI has defined a highly innovative approach in ETSI EN 303 146-4 based on a Radio Virtual Machine (RVM) concept. The RVM abstracts the Radio Application (RA) code generated with the ETSI-standardized programming interfaces in such a way that the software code can be executed directly (no middleware is required) on any hardware platform compliant with the ETSI software reconfiguration framework.

For software portability, Figure 7.6 illustrates a conceptual diagram showing how the RA code is abstracted through the RVM to be ported onto different hardware platforms. In this specific example, the RA code is made available to a multitude of different hardware platforms through the RVM. As shown on the right side of Figure 7.6, the RVM includes data objects (DOs) for data abstraction, abstract processing elements (APEs) for computational element abstraction, and abstract switch fabric (ASF) for switching the DOs and APEs. The RVM is indeed an abstract machine which abstracts the RA code for a given hardware platform. Therefore, the RVM allows the conversion of a given software component into a generic representation (as a result of front-end compilation), which is then optimized for the specific hardware resources available on a target platform (as a result of back-end compilation). Software developers are able to create software components without considering particular modem hardware details.

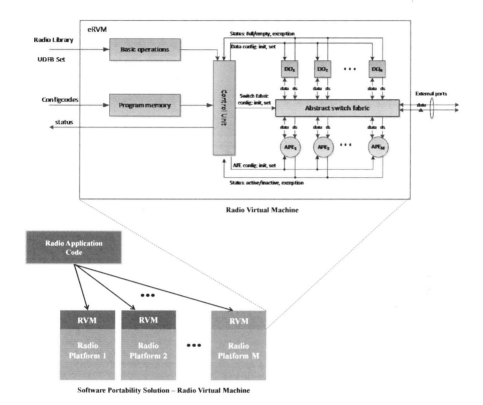

Figure 7.6: Concept of the radio virtual machine

This approach ensures code portability while maintaining efficiency. The latter is possible since no middleware is introduced and RadioApp designers have full flexibility for joint optimization of hardware and software designs. The radio virtual machine (RVM) is an abstract machine that is capable of executing configcodes and is independent of the hardware. The implementation of a RVM is target radio computer-specific and it has access to the back-end compiler (on the platform itself or externally) for just-in-time (JIT) or ahead-of-time (AOT) compilation of configcodes.

The RVM executes a particular algorithm presented as a data flow chart. In other words, the RVM is the result of replacing all operators and tokens in the particular data flow chart with abstract processing elements (APEs) and data objects (DOs), respectively. Each APE executes computations marked by the replaced operator identifier. These computations are taken from the radio library. Figure 7.7 illustrates a conceptual view of RVM processing. This process requires APE, DO, and radio library, that are defined as follows: APE abstracts a computational resource corresponding to the operation in a particular data flow chart. DO abstracts a memory resource where DO is an abstracted memory for storing data used during the procedure of radio processing. The reference/native radio library includes normative definitions and native implementation of all standard functional blocks (SFBs) for front-end and back-end compilation. Note that the computations included in the radio library are represented in terms of normative definitions or native implementations of SFBs, depending upon whether the radio library is used for front-end or back-end compilation, respectively.

We note that the user-defined functional blocks (UDFBs) are created through a combination of SFBs and represented as a data flow chart to be executed in the RVM. Alternatively, a UDFB is implemented as a stand-alone module or function, which can be mapped into one APE (this UDFB can be considered atomic), or into an eRVM/RVM (not atomic). UDFBs are not generally included into the radio library, but they are part of the radio application package.

The RVM begins to work immediately after the initialization of some DOs. All APEs execute asynchronous and concurrent computations. An individual APE executes the allocated operator if all the corresponding input DOs are full. APEs access DOs with operations "read," "read-erase," or "write." After reading input data from DOs, the APE executes the allocated operator and, if output DOs are empty, then the APE writes processed data. Any full output DO blocks the corresponding writing operation. The RVM executes computations until reaching the state when all APEs become inactive. In this state, there are not enough full DOs that can activate the inactive operators. The computations result in full DOs, which cannot activate the inactive operators.

Note that an output DO can become an input DO for a subsequent operator. This input DO can then activate the subsequent operator. Furthermore, the state or operation of a given APE is independent from the state of other APEs; each APE is atomic.

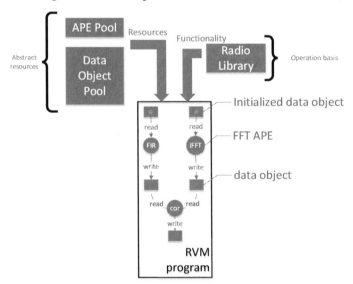

Figure 7.7: Conceptual diagram of radio virtual machine processing

In the sequel, the elementary RVM (eRVM) approach is described. An eRVM consists of components of basic operations, program memory, control unit (CU), abstract switch fabric (ASF), as well as APEs and DOs, that are defined as follows: eRVM does not contain another eRVM or RVM. Basic operations include operators either provided from the radio library as SFBs and/or from the UDFB set as UDFBs, which is mapped onto one single APE. Since UDFBs might be implemented as a stand-alone module/function which can be mapped onto one APE, in this case, basic operations include operators provided by the UDFB set and by the radio library as SFBs. Note that those UDFBs are atomic. For a RVM, the SFB or UDFB can be mapped onto an APE, RVM, or eRVM. In the eRVM case, the mapping to RVM or eRVM is not possible since it is the lowest level of hierarchy. Furthermore, from an execution perspective, there is no difference between SFBs and UDFBs.

The program memory is provided with configcodes, which determine the eRVM configuration. The CU generates initialization and set-up instructions for APEs, DOs, and ASF based on decoding configcodes stored in the program memory. The ASF connects APEs and DOs in accordance with CU signals. One DO can be connected with multiple APEs. One APE can be connected with multiple DOs. DOs from other eRVMs can be connected with ASF through external data ports. Figure 7.8 illustrates a block diagram of eRVM. Basic operations in eRVM consist of operations provided by the radio library and/or UDFB set. A target platform may or may not provide accelerators for some or all SFBs and/or UDFBs. Furthermore, three cases can be considered. First, RAP includes only SFBs; second, RAP includes only UDFBs; and third, RAP includes SFBs and UDFBs. Finally, and independent of the above, basic operations may include SFBs only, UDFBs only, or SFBs and UDFBs.

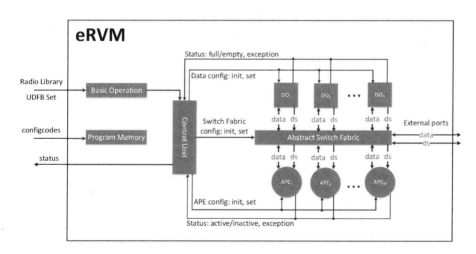

Figure 7.8: Elementary RVM

Figure 7.9 illustrates how a component of an existing chain (a transmission chain) is being replaced through an RVM-based software component. It is typically up to the manufacturer to decide which components of a target platform may be made available to a third-party software developer. In the past, it has proven difficult to fully open up a highly complex hardware environment; so, at the early stage of introducing the new technology, it may be preferable to enable such third parties to replace one or multiple of selected components by novel software components. Figure 7.9 illustrates the replacement of an abstract component B through a novel implementation by third-party software developers.

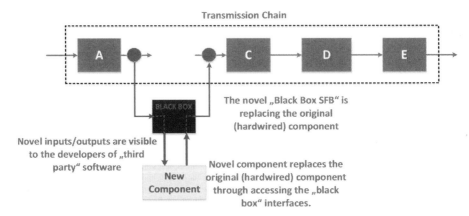

Figure 7.9: Example replacement of a component through interfacing with novel SFB provided by a third-party software provider

7.6 Multi-Radio Interface (MURI)

All interfaces required for the ETSI software reconfiguration framework are defined in ETSI EN 303 146-1/2/3/4. The multi-radio interface (MURI) defined in ETSI EN 303 146-1 is the most elementary interface and is responsible for basic features such as the provision of new software components, installation, execution, and so on. Further details about MURI are included in the sequel.

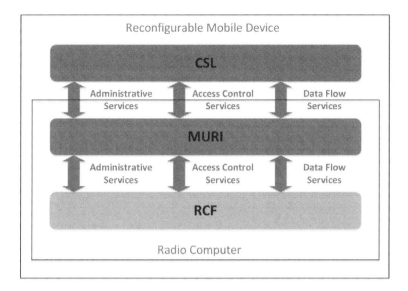

Figure 7.10: Interconnection between CSL and RCF using MURI for reconfigurable MD

As shown in Figure 7.10, MURI supports three kinds of services: administrative, access control, and data flow. Administrative services are used by some device configuration applications; the Administrator is included in the CSL and is used to (un)install a new unified radio application (URA) into the reconfigurable MD and to create or delete an instance of the URA. Installation and loading may both take place at device start-up time to set up the network connection and during run-time, whenever the reconfiguration of available URAs is needed. MURI does not make any assumptions on how and when the mobile device will detect the need of the reconfiguration. Access control services are used by the MPM to maintain the user policies and preferences related to the usage of different RATs and to make a selection between them. The modeling of such preferences and selection algorithms is not in the scope of the present document; however, the MURI specification covers the information exchange of RAT selection decisions between CSL and RCF. The preferences themselves may originate either locally from applications or end user settings, or in a distributed manner from a network operator or from a cognitive radio management framework. Data flow services are used by the networking stack of the reconfigurable MD, such as the TCP/IP stack. Data flow services represent the sets of (logical) link layer services, which are provided in a uniform manner regardless of which URAs are active. With these basic features, it is possible to enable the installation and activation of a unified radio application, to parameterize the configuration (by selecting a number of RATs to be operated simultaneously), to access data flow services, and to perform the deactivation and uninstallation of a unified radio application.

In a first-generation implementation of a software reconfigurable device, MURI alone may be implemented among the set of available interfaces. With this design choice, basic reconfiguration features will be available in the early stage. Over time, these basic features can be extended to cover further services. The design is thus future proof.

7.7 RRS Software Reconfigurability For Hardware Updates

The SW reconfigurability for gradual evolution in 8.2.5 may be necessary during the lifetime of the installed automotive RRS module. All the RF components—including up-down converter, power amplifier, filter, switch, diplexer, diversity switch, antenna tuner, and antenna—need to be changed when the covering frequency band is changed. Typically, a module manufacturer will design the RF system in such a way that all RF bands are covered which are expected to be relevant over the lifetime of the device. However, it would not be possible to predict all the required frequency bands for the next twelve years, which consists of the two years of the system development time and the ten years of the car lifetime. Therefore, the module manufacturer may want to design the RRS module to be RF-upgradable so that the customers can use the most up-to-date radio communication.

One possible implementation is making the automotive RRS module with plug-in RF modules. A new RF module can replace the existing RRS RF module, or a new RF module can be added to the other plug-in slot of the RRS module, keeping the existing RRS RF module. The latter would be more cost-effective at the extra initial RRS module cost for the extra plug-in slot. Both methods of upgrading the automotive RRS RF module would cost less than replacing the whole automotive non-RRS communication module, including all the baseband components. For the most up-to-date radio communication, the RRS RF module change may be necessary every five to ten years, but the frequency or cost of the RRS HW change would be drastically reduced compared to the non-RRS automotive communication module. If the most up-to-date radio communication is not necessary, the RRS HW change may not be necessary during the RRS module lifetime.

In the ETSI RRS system, even though the new frequency band is added by the RF module HW change, the baseband module HW does not need to be changed, but only the baseband module SW needs to be upgraded. The replaced or added RF module is also compatible with the reconfigurable radio frequency interface (RRFI) as defined in ETSI EN 303 146-2, so only the unified radio application (URA) needs to be upgraded by the RRS software reconfiguration to control the new RF components. Therefore, the ETSI RRS cannot upgrade the HW components by its nature; however, it is very HW upgrade-friendly.

7.8 The ETSI Software Reconfiguration Solution in Comparison to Other Alternatives

A multitude of software reconfiguration approaches exist. Among those, the Software Communications Architecture (SCA) is a very prominent solution. A short overview is given and the differences to the ETSI approach are outlined.

As mentioned in Software Communications Architecture Specification, Joint Tactical Networking Center (JTNC), August 2015, V4.1, the Joint Tactical Networking Center (JTNC) published the Software Communications Architecture (SCA). This architecture was developed to assist in the development of software-defined radio (SDR) communication systems and to capture the benefits of recent technology advances, which are expected to greatly enhance interoperability of communication systems and reduce development and deployment costs. As stated in Software Communications Architecture Specification User's Guide, Version: 4.1, a fundamental feature of the SCA is the separation of waveforms from the radio's operating environment. Establishing a standardized host environment for waveforms, regardless of other radio characteristics, enhances waveform portability. The waveform software is isolated from specific radio hardware or implementations by standardized APIs.

Nevertheless, the portability problem was not resolved within SCA; see, for example, Wireless Innovation Forum Top 10 Most Wanted Wireless Innovations and Acquiring and Sharing Knowledge for Developing SCA Based Waveforms on SDRs, Report RTO MP-IST-092 where it was pointed out as a primary problem of SCA. SCA (any version) middleware separates and isolates software from hardware, and therefore does not allow a joint optimization of hardware and software, which is the main source of efficiency for embedded devices. Although the middleware of SCA is quite sophisticated, it is too redundant and, thus, not efficient enough for commercial applications. The development of such middleware is quite costly for the civil industry. Historically, SCA was designed based on the distributed computing approach, but the modern terminals are built based on system-on-chips (SoC) where multiple intellectual property (IP) cores are integrated into a single chip. Still, since SoC-based technology does not assume distributed internal communications, it is not reasonable to support the baseline "client-server" model and sophisticated hardware agnostic transactions among software components. Meanwhile, since the "client-server" model proposed by the Object Management Group (OMG) is not formal, it cannot support a critical formal verification due to the software complexity for emerging multiple Radio Access Technology (multi-RAT) mobile devices working in a heterogeneous wireless network environment.

An alternative approach to the radio virtual machine (RVM) is the java virtual machine (JVM), which was considered in M. Gudaitis and J. Mitola III's article, "The Radio Virtual Machine" and Abdallah et al.'s "The Radio Virtual Machine: A Solution for SDR Portability and Platform Reconfigurability." The implementation aspects of the RVM were studied in Abdallah et al. as well as Tanguy Risset et al. in "Virtual

Machine for Software Defined Radio: Evaluating the Software VM Approach." But none of these sources treated the problem of RVM architecture efficiency. The authors rather focused on existing VMs based on the Von Neumann architecture, and evaluated corresponding overheads.

7.9   ETSI Security Framework for Software Reconfiguration

The ETSI security framework for software reconfiguration defined in ETSI TR 103 087 and TS 103 436 applies the traditional thinking of the confidentiality integrity availability paradigm of assuring proper behavior of the radio equipment, providing tools for secure deployment of the technology, assisting the users and developers in the avoidance of fraud, and supporting developers in proving conformance to the regulatory framework in which the equipment operates. Three assets are identified as the radio application, the reconfiguration policy that can be used in managing the reconfiguration of the equipment, and the Declaration of Conformity (DoC), a document with legal value.

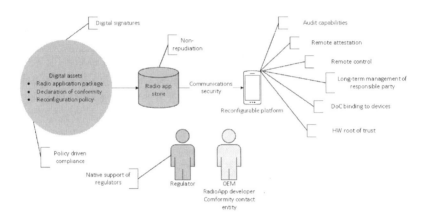

Figure 7.11: Security measures for the ETSI software reconfiguration framework

The role of security countermeasures is to defend the system against attack. The nature of attacks against a reconfigurable radio have been assumed to leverage the mutability of the platform where the wireless connectivity options of the platform are designed to be modifiable over time. A precursor for allowing radio applications to be installed is to have the base platform itself be secure to act as a root of trust and security. The rationale is to build on firm foundations (not to build on sand) and to make a strong binding of application to the root of trust, thus extending in-depth security through the evolving platform.

The installation of purposefully misbehaving radio applications and other malicious assets is among the greatest threats to the security of the radio equipment and of the users. There is a risk that legitimate radio applications are not used properly—for example, in the context of a user trying to bypass a hardware- or policy-based limitation, or in the context of device counterfeit. Additionally, an attacker may attempt to seize control of the equipment by taking advantage of a security vulnerability.

The overall architecture to achieving the security goals for RRS is that of a multiparty digital signature scheme complemented with a non-repudiation scheme, with entities in the system delivering cryptographically sealed and identified proof that actions have been taken to assure the operation of the value chain—for example, that conformance testing took place—and to make that proof available to authorized and trusted third parties. The result of the application of the above measures is an assurance that the platform and its applications will work securely against threats of manipulation, masquerading of any of the actors, and against regulatory bypasses.

The trusted third parties can be the equipment manufacturers and network operators. Regulatory bodies are natively supported as actors of the framework, which they can leverage to implement market surveillance and disturbance control.

Further extensions to the RRS security model have been developed that extend the scope of these proofs to allow for remote attestation of the radio by ensuring that only allowed radio applications exist on the RRS platform. Furthermore, they give high assurance of the correct behavior of radio applications. The model has provisions for a hardware root of trust, providing assurance of the software reconfiguration platform integrity to the highest possible industry standards. Remote control and long-term management features complement the model so that radio technology evolution and management are securely handled within the RRS framework.

To summarize, the security measures to address these threats in the ETSI software reconfiguration framework are

- the proof of the integrity of the radio applications, reconfiguration policy, and Declaration of Conformity
- the proof of the identity of the developer of radio applications, the issuer of the reconfiguration policy, and the issuer of the Declaration of Conformity
- the prevention of an asset installation when the asset is not provided by a legitimate actor
- the use of the reconfiguration feature as a security update mechanism
- the proof of conformance of the radio platform and radio application to the regulatory Declaration of Conformity, considering that the set of installed radio applications can change over time
- the prevention of illegitimate use of the Declaration of Conformity (in particular against counterfeit)
- the audit functionalities including a non-repudiation framework and remote attestation, the long-term management framework (e.g., transition of equipment responsibility from one manufacturer to another)

- the prevention of masquerade of stakeholders in the RRS value chain
- the prevention of code theft
- the supply chain integrity and assurance (which underpins all of the above measures)

7.10 Responsibility Management

With the development of a unified radio application, the radio behavior of the target devices typically changes, and the conformity to the application regulation must be revalidated. In Europe, Annex VII of the Radio Equipment Directive introduces a useful tool in order to update the applicable Declaration of Conformity of a radio equipment. In this context, radio virtual machine (protection) classes are introduced in order to find a trade-off between recertification efforts and base-band code development flexibility.

At one extreme of the RVM class, a high-level RVM class corresponds to full reconfigurability of the low-level parameters of an RVM, and necessitates a more extensive certification-testing process after the RVM has been reconfigured. At the other extreme of the RVM class, a low-level RVM class corresponds to a limited reconfigurability of the low-level parameters of an RVM. As the reconfigurability of the low-level parameters of this particular class of RVM is limited, a relatively less extensive certification-testing process is necessitated after the RVM has been reconfigured. Moreover, an RVM can have different RVM classes associated with different components of the RVM that relate to the reconfigurability of the low-level parameters of the respective components of the RVM.

Reconfiguration of an RVM of the highest-level RVM class may necessitate that the overall certification-testing process focuses on the certification of each reconfigured software component of the RVM. In such a situation, each respective reconfigured software component may need to be separately certified before one or more sets of reconfigured software components are certified together. For example, a reconfigured RVM software component "A" (for example, Wi-Fi) may need to be separately certified from a reconfigured and certified RVM software component "B" (for example, LTE). The certification process may then be such that the joint operation of separately certified reconfigured RVM software components "A" and "B" could then take place jointly.

At the other extreme of RVM classes, the lowest-level RVM class corresponds to a restricted reconfigurability of the low-level parameters of an RVM. For such a restricted level of reconfigurability, a developer of radio applications would only have limited access to the low-level parameters of an RVM. For example, the lowest-level RVM class would permit a radio application developer to have access to only the low-level parameters of the receiver chain of an RVM. Accordingly, the lowest level of RVM class would not need to utilize a corresponding detailed and thorough certification testing process because, for example, a radio platform operating a malfunctioning reconfigured RVM would not interfere with other radio platforms. Thus, the level of certification-testing for the lowest RVM class would be a less extensive certification-testing process than that used for the highest RVM class.

One or more medium- or intermediate-level RVM classes may also be established between the two extreme RVM classes that correspond to intermediate levels of reconfigurability of the low-level parameters of an RVM. An intermediate-level RVM class, for example, would allow more flexibility for reconfiguring low-level parameters of an RVM than the lowest-level RVM class, but would not permit the degree of reconfigurability that would be associated with the highest-level RVM class. Depending on the level of reconfigurability to the low-level parameters of an RVM, an intermediate-level RVM class may necessitate a certification-testing process for a compiled reconfigured RVM and underlying hardware that is more extensive than that corresponding to the lowest-level RVM class, but less extensive than that corresponding to the highest-level RVM class. For example, a certification based on the intermediate-level RVM software component might be obtained by contacting an authorized notified body and providing only a serial number for the RVM software component and an identification of the target device type on which the compiled reconfigured RVM would operate. In another example, there could be no requirement for a joint certification based on an RVM software component for a simultaneous operation with other RVM software components. That is, a certificate based on an RVM software component "A" (such as Wi-Fi) and a separate certificate based on another RVM software component "B" (LTE) would allow for a simultaneous operation of reconfigured components "A" and "B."

Another exemplary situation that may necessitate a relatively less extensive certification-testing process would be a radio application developer that only reconfigures non-transmission-related low-level parameters—for example, low-level parameters relating to a data interleave and/or a channel coder in the transmit/receive (TX/RX) chain of an RVM that otherwise has been defined to be of the highest-level RVM class. As nothing related to the spectral shaping of a transmitted signal is reconfigured by the reconfiguration of the data interleave and/or channel coder, a relatively less extensive certification-testing process could be used. Another exemplary situation would be a reconfiguration that involves changes targeting predefined frequency bands and/or bandwidths. In still other situations, there may be reconfigurations for which a certification-testing process may not be necessary.

In the context of software reconfiguration of radio parameters, a key question relates to responsibility: How to address the responsibilities of software and hardware manufacturers. Market surveillance authorities always request a single responsible party to be the contact point in case some malfunctioning equipment is identified. It is therefore challenging to clearly separate the responsibilities between hardware and software manufacturers—often, observed malfunctioning features cannot be easily traced back to issues in specific hardware or software components. The overall problem becomes less critical if all components (software and hardware) are being created by a single entity. In order to differentiate these two cases, the Telecommunications Conformity Assessment and Market Surveillance Committee (TCAM) has proposed a definition for two market models—horizontal and vertical markets. A vertical market comprises all hardware and software relating to the essential requirements (under the Radio Equipment Directive); all functions and underlying hardware—which are relevant for the Declaration of Conformity—are controlled by one single entity. A horizontal market includes independent companies placing hardware and SDR software (third-party software providers, etc.) separately on the market which, when used together, are subject to Declaration of Conformity with the essential requirements for the intended use of the equipment.

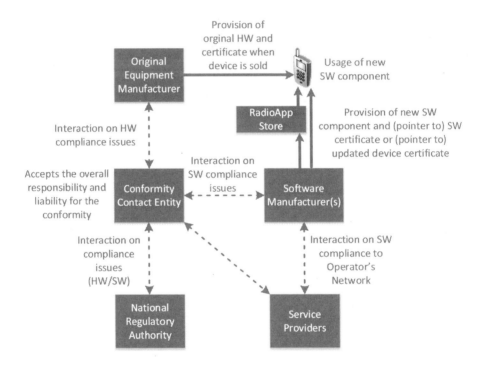

Figure 7.12: Alternate interactions in the horizontal market model with a single contact point

Responsibilities for the vertical market are quite straightforward. A single entity creates all software and hardware components and will sign the Declaration of Conformity (DoC) of the equipment and will thus take the overall responsibility and service as a contact point to market surveillance. The more complex horizontal market case has been discussed in ETSI's Reconfigurable Radio Systems (RRS) Technical Committee (TC). As a conclusion, TR 102 967 recommends the introduction of a "Conformity

Contact Entity" that takes the full responsibility of the entire equipment, including software and hardware components. This entity also serves as a contact to market surveillance.

When an issue occurs, the "Conformity Contact Entity" needs to further analyze the source of any potential problem and handle the interaction with hardware and software manufacturers internally, typically following a contractual agreement. In practice, the hardware manufacturer is also expected to take on the role of the "Conformity Contact Entity."

7.11 Conclusion

In future 5G communication systems, the usage of hardware and spectrum resources will be dramatically different from previous generations. Software reconfigurability solutions will allow chipset manufacturers to develop generic solutions that can then be tailored to a multitude of vertical applications, including vehicular applications among others. ETSI RRS has developed an entire reconfiguration ecosystem that provides solutions for technical, security, and conformity needs. The underlying solution is different from existing state-of-the-art software reconfiguration approaches in the sense that the manufacturer can choose any appropriate level of "openness" of the underlying platform (allowing for the replacement of an entire RAT or only specific components), and the radio virtual machine approach ensures a highly efficient approach that allows code portability without requiring the usage of an (inefficient) middleware.

For the vehicular communications application target, a first generation of products may only employ basic features of the ETSI RRS SW reconfiguration framework; in particular, the MURI interface is expected to be a suitable candidate that provides features such as downloading, installation, execution, and uninstallation of software packages onto a target vehicular platform.

In the next stage, vehicular applications may take full advantage of the software reconfiguration framework defined by ETSI EN 302 969, EN 303 095 and EN 303 146-1/2/3/4—in particular, offering a highly portable and yet efficient ecosystem for providing essential software components to vehicles. As a future vision, vehicle owners may indeed choose specific software components to add and upgrade automated driving-related features and multimedia applications. Those will be based on hardware components which are designed in a generic way and will be personalized through the upload of suitable SW components.

References

2/23/2016: Software Communications Architecture Specification User's Guide.

2010 IEEE 10th International Conference on Computer and Information Technology (CIT). Bradford, United Kingdom.

8/1/2015: Software Communications Architecture Specification, Joint Tactical Networking Center (JTNC).

Distributed Processing (IPDPS). Rome, Italy.

EN 302 969, 11/2014: Reconfigurable Radio Systems (RRS); Radio Reconfiguration Related Requirements for Mobile Device.

EN 303 095, 6/2015: Reconfigurable Radio Systems (RRS); Radio Reconfiguration Related Architecture for Mobile Devices.

EN 303 146-1, 11/2015: Reconfigurable Radio Systems (RRS); Mobile Device Information Models and Protocols; Part 1: Multi-radio Interface (MURI).

EN 303 146-2, 6/2016: Reconfigurable Radio Systems (RRS); Mobile Device (MD) Information Models and Pro-tocols; Part 2: Reconfigurable Radio Frequency Interface (RRFI).

EN 303 146-4, 4/2017: Reconfigurable Radio Systems (RRS); Mobile Device (MD) Information Models and Pro-tocols; Part 4: Radio Programming Interface (RPI).

ETSI TR, under development, Reconfigurable Radio Systems (RRS); Applicability of RRS with Existing Radio Access Technologies and Core Networks Security Aspects.

ETSI TS 103 436, 8/2016: Reconfigurable Radio Systems (RRS); Security Requirements for Reconfigurable Radios.

European Commission (11/2012): M/512, Standardisation Mandate to CEN, CENELEC AND ETSI for Reconfigurable Radio Systems.

European Commission (3/9/1999): Directive 1999/5/EC of the European Parliament and of the Council of 9 March 1999 on Radio Equipment and Telecommunications Terminal Equipment and the Mutual Recognition of their Conformity.

European Commission (4/16/2014): Directive 2014/53/EU of the European Parliament and of the Council of 16 April 2014 on the Harmonization of the Laws of the Member States Relating to the Making Available on the Market of Radio Equipment and Repealing Directive 1999/5/EC.

SDR Forum Workshop 2000.

Top 10 Most Wanted Wireless Innovations. Document WINNF-11-P-0014, Ver. V1.0.1. In: Wireless Innovation Forum 2011.

3GPP TR 102 967, 2015: Reconfigurable Radio Systems (RRS); Use Cases for Dynamic Equipment Reconfiguration.

3GPP TR 103 087, 2016: Reconfigurable Radio Systems (RRS); Security Related Use Cases and Threats in Reconfigurable Radio Systems.

3GPP TR 103 087, 6/2016: Reconfigurable Radio Systems (RRS); Security Related Use Cases and Threats in Reconfigu-rable Radio Systems.

3GPP TS 103 436, 2016: Reconfigurable Radio Systems (RRS); Security Requirements for Reconfigurable Radios.

Wireless Innovation Forum 2011.

Abdallah, Riadh Ben; Risset, Tanguy; Fraboulet, Antoine; Durand, Yves: The Radio Virtual Machine: A solution for SDR portability and platform reconfigurability. In: Distributed Processing (IPDPS). Rome, Italy, pp. 1–4.

Abdallah, Riadh Ben; Risset, Tanguy; Fraboulet, Antoine; Martin, Jérôme: Virtual Machine for Software Defined Radio: Evaluating the Software VM Approach. In: 2010 IEEE 10th International Conference on Computer and Information Technology (CIT). Bradford, United Kingdom, pp. 1970–1977.

Gudaitis, M.; Mitola III, J.: The Radio Virtual Machine. In: SDR Forum Workshop 2000.

Singh, S.; Adrat, M.; Antweiler, M.; Ulversoy, T.; Mjelde, T. M. O.; Hanssen, L. et al. (2013): *Acquiring and Sharing Knowledge for Developing SCA Based Waveforms on SDRs*. Fraunhofer FKIE.

Chapter 8
Outlook

The vehicle ecosystem is an industry in transformation. Automated and autonomous driving is accelerating the transformation of vehicles and the vehicle infrastructure from hardware-defined to software-defined solutions. The way toward automated and autonomous driving is forked with two developments running in parallel. There is the evolutionary improvement of private personal vehicles toward partial autonomy and there is the disruptive introduction of fully automated and autonomous driving as a new mode of mobility as a service is comprised of vehicle fleets. All global geos participating, but in different ways and at different speeds. In short and mid-term in China and the United States, the software giants of Baidu, Alibaba, Tencent, Google, Amazon and Uber are in the vanguard and in Europe, the premium vehicle manufacturers like Daimler, VW, Audi and PSA are in leadership but in the long term, fully automated vehicles and full autonomy will have the greatest impact in Asia. Vehicles become smartphones and servers on wheels.

Automated and autonomous driving vehicles are, at their early stages, trying to find their way into the vehicle ecosystem. The switch of responsibility from a human being as a driver at SAE level 0 to computing machines as a diver up to SAE level 5 is not easy and has to solve many issues in regulation, ethics and politics. Vehicle manufacturers advance from the current personal and private vehicles running in all environments with almost no location and application constraints with the driver in focus toward partial autonomy up to SAE level 3. The disruptors from the computing and communication ecosystem work at fully autonomic driving vehicles as part of huge vehicle fleets for mobility as a service, first to get implemented in urban environments under regulated and controlled conditions. There are large challenges to meet the required computing and V2X networking and connectivity performance, to maintain and increase the driver and passenger's trust in the technology, to overcome the heavily fragmented standardization and regulation landscape and to show up the predicted economic and environmental benefits.

Sensor technology like ultrasonic sound and video cameras have been used for many years in vehicles. The introduction of additional sensor from HD video to millimeter wave radar and LIDAR creates a huge performance demand for the processing platforms running the vehicle software stack as well as for the V2X networking and connectivity. Computing and communication become a limiting factor currently, including both hardware and software. Many use cases for evolved ADAS and automated and autonomous driving require sensor data fusion running continuously for example to create 3D scene models and HAD map updates in the vehicle as well as sensor data exchange with servers in the cloud. These HAD maps are very critical and need centimeter level accuracy and real-time updates for certain map layers. Applied deep learning for automated and autonomous driving classifies huge amounts of

DOI 10.1515/9781501507243-008

sensor data for the vehicle on board and cloud mapping data, which needs to be exchanged between vehicle and vehicle infrastructure.

Physical ECUs will continue to play a role in vehicles for some more years, even in automated and autonomous vehicles, since simple engine control and x-by-wire management systems are less expensive using traditional single-purpose embedded microprocessors and safety-critical dedicated systems. They are more reliable and function safer with dedicated hardware and software. But the functions including V2X networking and connectivity are more complex, and are developed more and more as software applications, since computing performance gets increased drastically in the vehicle computing platforms. So new and legacy features such as emissions control, adaptive cruise control, collision avoidance, navigation, route optimization are more and more implemented in software.

Many other ECUs, including those for ADAS vehicle operation or automated and autonomous driving functions get implemented at virtual ECUs on high performance server microprocessors with fast, multi-core processors and considerable amounts of on-board memory and a variety of in-vehicle and V2X networking connections to the vehicle ecosystem infrastructure. Solutions need to be developed for the purpose, not only for the required high performance processing of communication data but for reliability, predictability and functional safety.

Functional safety is the most challenging requirement for V2X networking and connectivity. System malfunctions are as important for functional safety as malicious attacks. But not all vehicle system issues stem from malicious acts. Securing a vehicle is about ensuring safe operation and protecting against malfunction, regardless of the cause. When there is a V2X networking and connectivity failure, the backup is to bring the vehicle to a safe state, and not to restore full functionality. Current vehicle functional safety is already complex, V2X networking and connectivity for automated and autonomous driving amplifies it. The current tools developed for functional safety need to be applied from the beginning of development until test and certification for V2X networking and connectivity as well. This includes hardware and software for encryption, real time intrusion detection and prevention systems, real-time countermeasures to mitigate attacks, such as isolating safety-critical V2X networking and connectivity systems and remote or cloud-based data logging and analysis.

Current vehicle standards for vehicle hardware and software architectures, for example SAE, Automotive Safety Integrity Level (ASIL), ISO 26262, AUTOSAR and GENIVI, evolve and attempt to support the opportunities of automated and autonomous driving in the vehicle ecosystem for all stakeholders. Since vehicle manufacturers need to collaborate with computing and communication stakeholders to shape the upcoming vehicle ecosystem together, these regulation and standardization bodies get backed by organizations like 5GAA. Nevertheless, new entrants like Google, Tesla, Uber, Alibaba or Tencent are still not there, since the software development on these new high-performance server platforms create plenty of opportunities for these software and web service specialists.

Today ECUs span over many different in-vehicle networks like CAN, LIN, MOST and FlexRay. Looking at the performance and efficiency challenges due to new sensor and computing technology on the vehicle data processing platforms an in-vehicle network evolution is required. Ethernet as a backbone is added for connecting ECUs. As we show, there are many opportunities for wireless vehicle communication. DSRC and cellular LTE and 5G NR compete for V2X networking and connectivity, but we make it very clear that automated and autonomous driving required technical solutions that are not yet augmented by V2X networking and connectivity, which is not part of the automated and autonomous real-time and functional safety core functionality. In the near term and midterm, automated and autonomous control use cases rely on on-board sensors, rather than on external communications.

Nevertheless, V2X networking and connectivity is for non-mission critical functions, for example use cases other than advanced ADAS and road safety. Examples for these non-safety use cases are mobile as a service business cases or infotainment. V2X networking and connectivity is used for non-safety, mobile as a service use case as of direct infotainment communications between two vehicles, communications between vehicle and surrounding infrastructure for offline data exchange and processing or communications between vehicle and services that run in the cloud such as fleet management, system performance and status reporting. Vehicle-to-cloud uses cellular LTE and 5G NR. For communications between the vehicle and surrounding infrastructure which is very dynamic in time and space it needs to be seen if 5G NR comes up with efficient solutions.

There is no doubt that the mass deployment of advanced autonomous vehicle features is imminent. Security services will dramatically decrease the number of road accidents and fatalities. At the same time, road management will become more efficient, less space will be required for road infrastructure, and traffic congestions and overall emission levels will be reduced. Autonomous driving will gradually be introduced, as an early stage technology is already in place. Vehicular communication will support corresponding features and make the vehicle a safer place to be!

Looking further into the future, we may even anticipate a fully automated vehicular ecosystem in which the human being is reduced to an undesired element, introducing instability into an otherwise perfect chain. In such a world, there will no longer be a driving wheel in vehicles. Humans will not make any decisions within this ecosystem, since only machines will be able to execute traffic rules by the letter and to consume multi-dimensional sensor data in real-time to make the optimum decision in any given situation. This picture may sound frightening, but we believe that we shouldn't worry just yet—there is still a long way to go and an infinite number of obstacles to overcome to make this vision a reality.

In the short-term, we have the rather pragmatic problem of introducing new features into the market and to obtain general acceptance. A specific problem relates to the competition between IEEE 802.11p-based DSRC and LTE C-V2X vehicular communication standards. It is widely expected that IEEE 802.11p-based DSRC will be the first entrant to the market, while LTE C-V2X will follow a couple of years later. Taking a typical renewal cycle time of vehicles into account (approximately ten years), a key question of the industry is: Which market penetration will be achieved by which vehicle communications standard?

Here, several possible scenarios are envisioned. One is that IEEE 802.11p-based DSRC takes all the market in the early stage, starting in 2018. The reason is that LTE C-V2X products are not yet commercially available. Contrary to LTE C-V2X, which will transition to 5G vehicular services, there is no evolution path for IEEE 802.11p-based DSRC. The obvious conclusion is that LTE C-V2X will gradually take over the market until IEEE 802.11p-based DSRC has disappeared entirely. This market behavior is illustrated below.

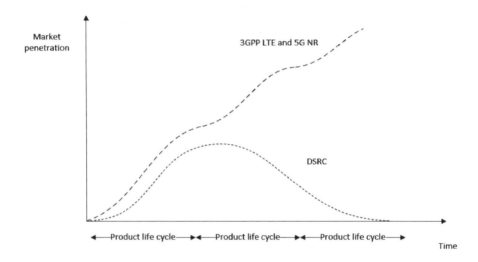

Figure 8.1: DSRC disappears

Another possibility is that we will see both LTE C-V2X and IEEE 802.11p based-DSRC prevail. One of the arguments in favour of such a scenario is that a massively deployed first-generation technology—such as IEEE 802.11p-based DSRC, in our case) is difficult to be removed from the market. A typical example is 2G cellular technology, which continues to survive even while we are transitioning from 4G to 5G technology. The corresponding market behavior is illustrated below.

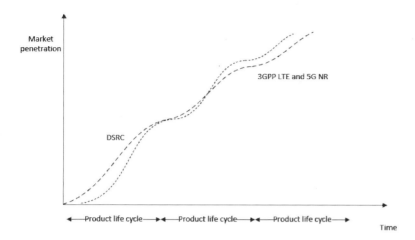

Figure 8.2: Both, DSRC and LTE and 5G NR get market acceptance

Finally, due to its early stage mass introduction, IEEE 802.11p-based DSRC may take and retain the market for basic safety features for the foreseeable future. In such a scenario, LTE C-V2X is likely to take the role of an "on the top" system by providing additional services that are typically limited to vehicles in the upper price segment. Consequently, the market penetration will lag behind the status of IEEE 802.11p-based DSRC. The corresponding market behavior is illustrated in Figure 8.3.

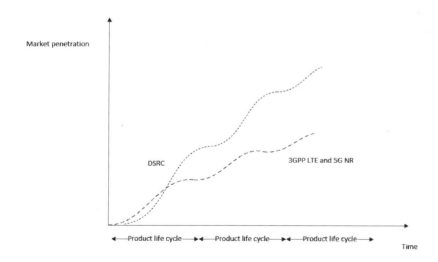

Figure 8.3: DSRC lead market penetration scenario

The final level of market acceptance for IEEE 802.11p-based DSRC and LTE C-V2X is hard to anticipate, so we need to observe the evolution. For the drivers and passengers, however, the existence of two competing systems is a nuisance. It basically leads to additional cost on one hand, with little to no added value on the other side. It is preferable to see a single system prevail. Due to the static nature of IEEE 802.11p-based DSRC and the clear evolution path of LTE C-V2X toward 5G technology and beyond, the gold option is to have LTE C-V2X prevail.

The market uncertainties outlined above, however, lead to challenges with respect to upcoming infrastructure investments. Corresponding political and industrial stakeholders are expected to contemplate the following questions.

- Should we delay the network deployment until we see the market stabilize? Will 3GPP LTE V2X, IEEE 802.11p-based DSRC, or both prevail in the long term?
- Should we bet on 3GPP LTE V2X only? A main argument for such a decision includes the fact that it is the technology with a clear evolutionary path toward 5G technology and beyond.
- Should we bet on IEEE 802.11p-based DSRC only? A main argument for such a decision includes the fact that this technology will be deployed on a large scale a number of years earlier than the 3GPP LTE V2X solution and will dominate the market, at least in the early stage.
- Should we deploy both 3GPP LTE V2X and IEEE 802.11p-based DSRC? In the case that only one of the two technologies prevail, the respective other investment may be lost in the mid- to long-term.

The recommendation of the authors is to introduce a flexible, software-reconfigurable solution that can be tailored to support 3GPP LTE V2X, IEEE 802.11p-based DSRC, or both. As illustrated below, a minimum (hard-wired) core of both technologies is proposed to be combined with a software-reconfigurable core consisting of programmable components such as digital signal processors (DSPs), field programmable gate arrays (FPGAs), and so on. Furthermore, additional required components, such as memory, are added. The flexible resources can be easily assigned to either or both of the vehicular communication technologies. It will also be possible to adapt the allocation of resources to the market penetration of the respective technologies. More market penetration will typically lead to higher traffic levels and more processing power. In case one technology finally prevails; the corresponding resources are correspondingly reallocated. The basic principle is illustrated in Figure 8.4 for any V2X-capable RSU or eNB entity.

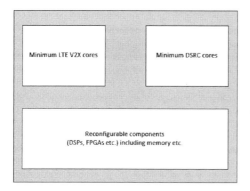

Figure 8.4: Adaptive V2X networking and connectivity system

The upper proposals lead to a reasonable business case independent of either of the two technologies, 3GPP LTE V2X and IEEE 802.11p-based DSRC, prevailing. There is an overall agreement across the vehicle ecosystem stakeholders that upcoming autonomous vehicles require a new internal networking and connectivity system with higher data rates, increased security, and lower costs. Evolved Ethernet with augmented wireless connectivity plays an important role here.

V2X wireless networking and connectivity augments the core functionality of vehicles, but is not part of it. Vehicle autonomy is improved by V2X wireless networking and connectivity but does not particularly rely on it for real-time functions. Various wireless standards, such as DSRC and cellular, compete for V2X inside and outside the vehicle. DSRC will most likely be mandated in the United States, but adoption is going to take off slowly, whereas DSRC will be explored in Europe and Asia. All regions will see an assessment of cellular V2X either as a back-up or first-priority, whereas vehicle-to-cloud will certainly use cellular networking and connectivity.

References

2014 IEEE 28th International Conference on Advanced Information Networking and Applications (2014). 2014 IEEE 28th International Conference on Advanced Information Networking and Applications (AINA). Victoria, BC, Canada, 13.05.2014 - 16.05.2014: IEEE.

2016 IEEE 19th International Conference on Intelligent Transportation Systems (ITSC) (2016). 2016 IEEE 19th International Conference on Intelligent Transportation Systems (ITSC). Rio de Janeiro, Brazil, 01.11.2016 - 04.11.2016: IEEE.

2016 IEEE International Conference on Communications (ICC). 22-27 May 2016 in Kuala Lumpur, Malaysia (2016). ICC 2016 - 2016 IEEE International Conference on Communications. Kuala Lumpur, Malaysia. Piscataway, NJ: IEEE.

BITKOM (2017): Car2x-Kommunikation - Technologieneutralität im 5.9 GHz Band. BITKOM.

Intel Corporation (9/1/2016): Internet of things: the future of autonomous driving.

National Highway Traffic Safety Administration; U.S. Department of Transportation (2016): Fact
    Sheet: Improving Safety and Mobility Through Vehicle-to-Vehicle Communication Technology.
    National Highway Traffic Safety Administration; U.S. Department of Transportation.
Bizon, Nicu; Dascalescu, Lucian; Mahdavi Tabatabaei, Naser (2014): Autonomous vehicles. Intelli-
    gent transport systems and smart technologies. Hauppauge, New York: Nova Science Publish-
    ers, Inc (Engineering tools, techniques and tables).
Demler, Mike; Case, Loyd; Gwennap, Linley (2016): A Guide to Processors for Advanced Automotive,
    The Linley Group.
Dr. Meng Lu (2016): Evaluation of Intelligent Road Transport Systems: Methods and Results. IET.
Kandel, Paul; Mackey, Matt (5/25/2016): The road to autonomous driving.

Appendix A

1.1 List of acronyms

3GPP	3rd Generation Partnership Project
5G	5th Generation (Communication System)
5GAA	5G Automotive Association
5G-PPP	5G Infrastructure Public Private Partnership
AACVTE	Ann Arbor Connected Vehicle Test Environment
AASHTO	American association of state highway officials
ABS	Anti-lock Braking System or Anti-Skid Braking System
AC	Airbag Control
ACC	Adaptive/Automatic Cruise Control
ACEA	European Automobile Manufacturers' Association
ADAS	Advanced Driver Assistance Systems
ADASIS	Advancing Map-Enhanced Driver Assistance Systems
ADC	Automatic Distance Control
AEB	Auto emergency braking
AFS	Adaptive Front Lighting Systems
AI	Artificial Intelligence
AOT	Ahead-of-Time
APA	Active Park Assist
APE	Abstract Processing Element
AR	Augmented Reality
ARQ	Automatic Repeat Request
ASF	Administrator Security Function
ASIC	Application-Specific Integrated Circuit
ASIL	Automotive Safety Integrity Levels
ASN.1	Abstract Syntax Notation One
AUTOSAR	Automotive Open System Architecture
AVB	Audio Video Bridging over Ethernet
BASt	Bundesanstalt für Straßenwesen (German Government Services for Street Operation)
BE	Backend (Compiler)
BER	Bit Error Rate
BLE	Bluetooth Low Energy
BLER	Block Error Rate
BS	Base Station
BSW	Blind Spot Warning
C-ITS	China ITS Industry Alliance
C-V2X	Cellular Vehicle-to-Everything (Communication)

DOI 10.1515/9781501507243-009

C2C	Car-to-Car
C2C-CC	CAR 2 CAR Communication Consortium
CA	Cooperative Awareness
CAM	Cooperative Awareness Message
CAN	Controller Area Network
CCSA	China Communications Standards Association
CCU	Communication Control Unit
CEC	Centralized and Domain-oriented Electronic Control
CEPT	Conférence Européenne des Administrations des Postes et des Télécommunications
CITS	Cooperative Intelligent Transportation Systems
CLC	Cooperative Lane Change
CLEPA	European Association of Automotive Suppliers
CM	(RRS) Configuration Manager
CNN	Convolutional Neural Networks
CoCA	Cooperative Collision Avoidance
CP	Cooperative Perception
CR	Cognitive Radio
CSL	Communication Services Layer
CSMA	Carrier Sense Multiple Access
CSMA/CA	Carrier Sense Multiple Access/Collision Avoidance
CSMA/CD	Carrier Sense Multiple Access/Collision Detection
CSPD-COM	Curve Speed Compliance
D2D	Device-to-Device (Communication)
DENM	Decentralized Environmental Notification Message
DFKI	Deutsche Forschungszentrum für Künstliche Intelligenz (German Research Center for Artificial Intelligence)
DL	Downlink
DNN	Deep Neural Networks
DNPW	Do not pass Warning
DO	Data Object
DoC	Declaration of Conformity
DOT	U.S. Department of Transportation
DSP	Digital Signal Processor
DSRC	Dedicated Short-Range Communication
E2E	End-to-End
EATA	European Automotive and Telecom Alliance
ECC	Electronic Communications Committee
ECTA	European Chemical Transport Association
ECU	Electronic Control Units
EEBL	Emergency Electronic Brake Light
eMBB	Enhanced Mobile Broadband

eMBMS	Enhanced Multimedia Broadcast and Multicast Service
eMTC	Massive Machine-Type Communications
eNB	Evolved Node B (Base Station)
EPS	Electric Power Steering
eRVM	elementary Radio Virtual Machine
ESP	Electronic Stability Control
ETC	Electronic Toll Collection
ETNO	European Telecommunications Network Operators' Association
ETSI	European Telecommunications Standards Institute
EVACINFO	Emergency Communications and Evacuation Information
FAAS	Freight-as-a-Service
FC	Flow Controller
FCC	Federal Communications Commission
FCW	Forward Collision Warning
FD	Flexible Data-Rate
FDM	Frequency Division Multiplexing
FFT	Fast Fourier Transform
FHWA	Federal Highway Administration
FOTA	Firmware Over-the-Air
FPGA	Field Programmable Gate Array
FRMWUPD	OTA Firmware Update
GLONASS	Globalnaja nawigazionnaja sputnikowaja sistema, englisch: Global Navigation Satellite System
GN	GeoNetworking
GNSS	Global Navigation Satellite System
GPS	Global Positioning System
GPU	Graphic Processing Units
GSA	Global mobile Suppliers Association
GSM	Global System for Mobile Communications
GSMA	GSM (Global System for Mobile Communications) Association
HARQ	Hybrid Automatic Repeat Request
HAV	Highly Automated Vehicles
HEV	Hybrid Electric Vehicle
HD	High Definition
HEVC	High Efficiency Video Coding
HMI	Human–Machine Interface
HVAC	Heating, Ventilation and Air Conditioning
HW	Hardware
I-SIGCVDAT	Data for intelligent Traffic Signal System
ICW	Intersection Collision Warning
IEEE	Institute of Electrical and Electronics Engineers
IETF	Internet Engineering Task Force

IMA	Intersection Movement Assist
IMT	International Mobile Telecommunications
InC	Within Network Coverage
IoT	Internet of Things
IP	Internet Protocol
IPv6	Internet Protocol Version 6
ISIC	Inter Cell Interference Coordination
ISO	International Organization for Standardization
ITS	Intelligent Transportation Systems
IVI	In-Vehicle Infotainment
JIT	Just-in-Time
LCA	Lane Change Warning and Assist
LCW	Lane Change Warning
LDW	Lane Departure Warning
LIDAR	Light Detection and Ranging
LIN	Local Interconnect Network
LLC	Logical Link Control
LOS	Line-of-Sight
LTA	Left Turn Assist
LTE	Long Term Evolution (4th Generation Communication System)
LVDS	Low-Voltage Differential Signaling
MAAS	Mobility-as-a-Service
MAC	Medium Access Control (Layer)
MAP	Map Data
MCD	Multimedia Content Dissemination
MCU	Micro Controller Units
MD	Mobile Device
MDRC	Mobile Device Reconfiguration Classes
MIMO	Multiple Input Multiple Output
MOST	Media Oriented Systems Transport
MPM	Mobility Policy Manager
MRC	Multiradio Controller
MURI	Multi Radio Interface
NCU	Navigation Control Unit
NFC	Near Field Communication
NFV	Network Function Virtualization
NGMN	Next Generation Mobile Networks
NGN	Next Generation Network
NHTSA	National Highway Traffic Safety Administration
NLOS	Non-Line-of-Sight
NPRM	Notice of Proposed Rulemaking
NR	New Radio

O&M	Operations and Maintenance
OEM	Original Equipment Manufacturer
OoC	Out of Network Coverage
OS	Operating System
OSI	Open Systems Interconnection Model
OTA	Over-the-Air
OVC	Oversize Vehicle Compliance
P2I	Pedestrian-to-Infrastructure
PARMLD	Parameter Uploading and Downloading
PCW	Pre-Crash Warning
PD	Pedestrian Protection
PDR	Packet Delivery Ratio
PEDINXWALK	Pedestrian in signalized Intersection Warning
PED-SIG	Mobile visually impaired Pedestrian Signal System
PHY	Physical (Layer)
PKI	Public Key Infrastructure
PLMN	Public Land Mobile Network
PN	Partial Networking
PRACH	Physical Random Access Channel
ProSe	Proximity Services
PSBCH	Physical Sidelink Broadcast Channel
PSCCH	Physical Sidelink Control Channel
PSSCH	Physical Sidelink Shared Channel
QAM	Quadrature Amplitude Modulation
QoS	Quality-of-Service
R&TTED	Radio and Telecommunication Terminal Equipment Directive
RA	Radio Application
RadioApp	Radio Application
RAP	Radio Application Package
RAT	Radio Access Technology
RCF	Radio Control Framework
RCM	Radio Connection Manager
RCTA	Rear Cross Traffic Alert
RDS	Radio Data System
RED	Radio Equipment Directive
RFMON	RF monitoring
RIOH	Research Institute of Highway
RKE	Remote Keyless Entry (Systems)
RLVW	Red Light Violation Warning
RM	Resource Manager
RNN	Recurrent Neural Networks
ROI	Regions of Interest

ROS	Radio Operating System
RPI	Radio Programming Interface
RRFI	Reconfigurable Radio Frequency Interface
RRS	Reconfigurable Radio System
RSSI	Received Signal Strength Indication
RSU	Road Side Unit
RTE	Runtime Environment
RTC	Regional Transportation Commission
RVM	Radio Virtual Machine
RX	Receiver
SAE	U.S. Society of Automotive Engineers
SC-FDM	A Single Carrier Frequency Division Multiple Access
SDO	Standards developing Organization
SDR	Software Defined Radio
SFB	Standard Functional Block
SI	Study Item
SLSS	Sidelink Synchronization Signals
SOTA	Software Over-the-Air
SPAT	Signal Phase and Time
SPD-COM	Speed Compliance
SPDCOMPWZ	Speed Compliance Work Zone
StVG	German Road Traffic Law
SVW	Stationary Vehicle Warning
SW	Software
TB	Terabytes
TC	Technical Committee
TCAM	Telecommunications Conformity Assessment and Market Surveillance Committee
TCP	Transmission Control Protocol
TCU	Transmission Control Unit, Telematics Control Unit
TDC	Traffic Data Collection
TDMA	Time Division Multiple Access
TEE	Trusted Execution Environment
TeSo	Tele-operated Support
TIAA	Telematics Industry Application Alliance
TJW	Traffic Jam Warning
TLI	Traffic Light Information
TMC	Traffic Message Channel
ToF	Time of Flight
TPMS	Tire Pressure Monitoring System
TR	Technical Report
TS	Technical Specification

TSN	Time Sensitive Networking
TSR	Traffic sign recognition
TTI	Transmission Time Interval
TTS	Traffic Technology Services
TÜV	Technischer Überwachungsverein (German businesses that provide inspection and product certification services)
TX	Transmitter
UDFB	User-Defined Functional Block
UDP	User Datagram Protocol
UE	User Equipment
UI	User Interface
UICC	Universal Integrated-Circuit Card
UL	Uplink
UMTS	Universal Mobile Telecommunications System (3rd Generation Communication System)
URA	Unified Radio Application
URAI	Unified Radio Application Interface
URLLC	Ultra-Reliable and Low Latency Communications
USB	Universal Serial Bus
UTRAN	UMTS Terrestrial Radio Access Network
VaaS	Vehicle-as-a-Service
VaD	Video Data Sharing for assisted and improved Automated Driving
V2C2I	Vehicle-to-Cloud-to-Infrastructure
V2G	Vehicle-to-Grid
V2H	Vehicle-to-Home
V2I	Vehicle-to-Infrastructure
V2N	Vehicle-to-Network
V2P	Vehicle-to-Person
VCU	Vehicle Control Unit
VRU	Vulnerable Road User
VTRW	Vehicles turning right in Front of Buses
V2V	Vehicle-to-Vehicle
V2X	Vehicle-to-Everything
VDA	German Association of the Automotive Industry
VR	Virtual Reality
WAVE	Wireless Access in Vehicular Environments
WI	Working Item
Wi-Fi	Wireless Fidelity
Wi-Gig	Wireless Gigabit

Index

802.11p-based DSRC 121, 123, 125, 132, 197, 198, 199, 200, 201

A
AACVTE (Ann Arbor Connected Vehicle Test Environment) 41
ABI Research 56, 117
ABS 91, 92
AC (airbag control) 52, 91, 203
Acceleration 4, 36, 62, 65, 91, 111
Access control services 185
Accidents 6, 7, 8, 19, 20, 21, 22, 69, 70
ACEA 33, 117, 119
Actuators 18, 89, 91, 96, 159
Adaptive cruise control 4, 42, 47, 91, 93, 95, 98, 156, 158
Adaptive cruise control (ACC) 11, 31, 42, 52, 54, 91, 95, 98, 156
Adaptive drivetrain management 30
ADAS (advanced driver assistance systems) 8, 24, 42, 44, 97, 98, 106, 157, 158
ADAS applications 44, 47, 164
ADAS systems 90, 112
ADASIS Advancing Map-Enhanced Driver Assistance Systems 203
Administrator 177, 178, 185
Advanced information networking 149, 150, 201
Advanced information networking and applications. *See* AINA
AEB (Auto emergency braking) 95, 203
AINA (Advanced Information Networking and Applications) 149, 150, 201
Airbag control (AC) 52, 91, 203
Alarms 41, 42
Alerts 37, 46, 121
Allocation 81, 84
Americas 41, 54, 56
Android auto 24, 154, 156, 164
Angles, horizontal aperture 93, 94
Ann Arbor connected vehicle test environment (AACVTE) 41
Annex VII 171, 190
Antenna 174, 177, 186
Anticipation, driver's 4, 5

APEs (abstract processing elements) 180, 181, 182, 183
Appendix 203, 204, 206, 208
Application note 88
Application processor 174
Application types 125
Applications 9, 16, 25, 35, 43, 56, 78, 122, 128
– multi-radio 174, 178
– non-safety 63, 101, 126, 179
– vehicular 85, 131, 132, 193
– vertical 55, 129, 193
Apps 115, 125, 152, 155, 156, 161, 163, 164, 174
Architecture 106, 110, 158, 160, 174, 179, 187, 189, 194
– software communications 169, 187
– station reference 127, 128
Area controllers 97
Artificial intelligence 1, 6, 17, 48, 61, 65, 72, 203, 204
ASF (administrator security function) 178, 180, 183
ASIL (automotive safety integrity levels) 16, 79, 111, 196, 203
Assist, intersection movement 33, 37, 44, 46, 52
Assistance, lateral 73
Assistance systems 28, 42, 62, 91
– advanced driver 8, 24, 97, 154, 157
ATM 124, 125
Audi 16, 23, 25, 36, 37, 41, 154, 155, 157
Augmented reality (AR) 37, 72, 130, 153, 203
Authentication 35, 71, 126
Auto emergency braking (AEB) 95, 203
Automated driving 26, 56, 117, 118, 165
Automated driving roadmap 25
Automated vehicles 6, 7, 17, 18, 22, 25, 27, 61, 114
Automatic distance control (ADC) 43
Automation 4, 5, 17, 22, 51, 60, 61, 62, 73
– conditional 4, 5, 62, 73
– high 4, 5, 62, 74
– partial 4, 62, 73
Automation level 2, 48, 49, 51, 55, 61, 164

DOI 10.1515/9781501507243-010

Automotive association 82, 87, 149, 203
Automotive user interfaces 56, 118, 165
Autonomous driving 20, 26, 56, 57, 118
Autonomous vehicles 2, 3, 14, 18, 28, 59, 72, 108, 118
AUTOSAR 106, 107, 164, 196
Available online 116, 149
Awareness, road hazard 71, 72

B
Band 34, 81, 82, 84, 85, 100, 139
Bandwidth 3, 99, 147, 158, 160, 191
Bird's eye 34, 35, 36, 49, 52, 158
BITKOM 201
Black box 52, 184
BLE (Bluetooth low energy) 59, 160, 203
Blind spot warning (BSW) 31, 33, 38, 52, 203
Blocks 55, 112, 113, 182
Bluetooth 16, 59, 86, 107, 112, 153, 154, 160, 162
Blue-tooth 152, 161
Bluetooth for in-car infotainment systems 165
Bluetooth low energy (BLE) 59, 160, 203
BMW 17, 36, 44, 45, 61, 65, 155, 157
Brakes 11, 37, 42, 55, 89, 91, 100
Bundesanstalt für Straßenwesen (BASt) 4, 51, 203
Business cases 29, 51, 52, 53, 54, 201
Bytes 66, 69, 86, 128, 135

C
CAM (Cooperative awareness messages) 14, 67, 101, 102, 121, 128
Cameras 8, 43, 44, 53, 72, 93, 153, 157, 158
Capabilities 3, 7, 42, 48, 89, 92, 129, 130, 172
– vehicle on-board sensors 65
Cases 29, 41, 42, 45, 49, 54, 55, 62, 63
CCP (consecutive CAM period) 63
CCSA (China Communications Standards As-sociation) 38
CCU (communication control unit) 152
Cell 15, 102, 103, 131
Cellular LTE 86, 103, 197
Certificate 24, 191, 192
Certification 16, 21, 170, 172, 173, 190, 191, 196

Certification-testing process 190, 191
– extensive 190, 191
CH 81
Channels 13, 75, 81, 82, 84, 141
China communications standards associa-tion (CCSA) 38
C-ITS 38, 39, 46, 78, 87, 115, 117
Classes, mobile device reconfiguration 172, 173
CLC (cooperative lane change) 39, 50, 204
Cloud 12, 43, 47, 48, 50, 72, 89, 90, 99
Cloud data centers 14, 72, 73
CNN (convolutional neural networks) 99, 106, 204
Code efficiency 180, 181, 183
Collision 42, 68, 92, 101
Collision avoidance 7, 8, 9, 11, 17, 71, 93, 196
– cooperative 34, 35, 36, 39, 44
Communications links 6, 15, 131, 138, 139
Communications protocols 64, 160
Communications requirements 9, 23, 24, 61, 68, 95
Communications services 9, 54, 128, 138, 139, 178
Communications solutions 36, 48, 158, 164
Communications stakeholders 23, 106, 196
Communication system 132, 187, 193, 203, 209
Communications technologies 23, 38, 49, 63, 78, 90, 114, 153, 155
Compilation, back-end 172, 180, 182
Compiler 175, 176, 203
Compliance 16, 80, 170, 171, 188
Components 8, 9, 24, 25, 71, 126, 169, 177, 184
Computations 181, 182
Computer 1, 2, 18, 21, 154
Computing 2, 3, 10, 23, 27, 89, 108, 155, 195
– mobile-edge 149, 150
Computing performance 96, 98, 196
Computing platforms 90, 96, 97, 98, 99, 106, 113, 114
Configcodes 175, 176, 181, 183
Configuration manager 174, 177, 178, 204
Conformity 169, 171, 188, 189, 190, 192, 193, 194
– new Declaration of 170, 171

Conformity contact entity 188, 192, 193
Connected vehicle system diagram 122, 123
Connected vehicles 15, 25, 28, 59, 113, 114, 150, 158, 160
– networked 25
Connected vehicles system 165
Connections 160, 161
Connectivity 13, 15, 17, 23, 54, 55, 59, 82, 114
– embedded 102, 161
– intermittent 59, 86
Connectivity next-generation technologies 100
Connectivity regulation 79, 81, 83
Connectivity requirements 28, 66, 69, 70
Connectivity solutions 24, 62, 87
Connectivity standards 73, 75, 77, 86, 87, 115, 165
Connectivity system 86, 112, 113, 196, 201
Connectivity technologies 71, 73, 80, 81, 82, 86, 89, 100, 161
– multiple vehicle 157
Connectivity technology candidates 64
Consecutive CAM period (CCP) 63
Context 78, 80, 81, 167, 169, 171, 189, 190, 191
Control 1, 17, 19, 20, 74, 80, 116, 156, 164
– longitudinal 4, 5
– remote 41, 46, 70, 71, 188
– transverse 4, 5
Control functions 115, 163
Control loops 70, 127
Control system 100, 151, 162, 163, 164
Control unit (CU) 183
Cooperative adaptive cruise control 31, 32, 41, 52, 54, 55, 68, 74, 78
Cooperative lane change (CLC) 39, 50, 204
Cooperative vehicle-highway automation system 30, 31, 32, 52
Core network 133, 134
Cost 1, 71, 72, 91, 122, 167, 186
Countries 22, 27, 28, 164
Coverage 6, 8, 25, 39, 63, 72, 103, 105, 136
CP (cooperative perception) 77, 78, 95, 204
Crashes 122, 124
CSL (communication services layer) 174, 177, 178, 179, 185
CU (control unit) 183
C-V2X 29, 36, 38, 49, 87, 88

C-V2X evolution 49, 143, 145, 147, 149
C-V2X Phase 130, 131, 133, 135, 137, 139, 141

D
Data centers 3, 48, 61, 65, 72, 90, 91, 98
Data exchange 9, 43, 66, 74, 101
Data flow chart 181, 182
Data flow services 185
Data frames 74, 75, 101
Data objects 180, 181, 182, 204
Data protection 19, 21
Data throughput 24, 28, 35, 95, 116
DCC (Decentralized Congestion Control) 77, 78, 126

Decision 1, 2, 18, 20, 106, 108, 124, 125, 200
Dedicated short-range communications 13, 22, 36, 56, 88, 118
Deep neural networks (DNN) 99
Degrees 4, 5, 92, 93, 94, 163, 191
DENM (decentralized environmental notification messages) 14, 101, 121, 128
Deploy 38, 87, 124, 125, 200
Deployment 5, 6, 33, 72, 75, 117, 122, 124, 128
Design 15, 78, 113, 116, 117, 125, 130, 186
Destination application 35
Detection 12, 16, 48, 52, 92
Detection range 93, 94
Development 17, 18, 73, 74, 89, 90, 107, 153, 187
Development phases 112, 130
Devices 101, 102, 103, 155, 168, 170, 171, 185, 186
– output 154, 163
– target 167, 174, 190
Diagnostic data 16, 31
Diagnostics 100, 152, 157
Digital signal processors (DSPs) 172, 173, 200
Distance 4, 63, 91, 92, 93, 94, 114
DL 62, 102, 146, 204
DNN (Deep neural networks) 99, 204
DOI 88, 89, 117, 118, 121, 150, 151, 165, 167
Domain controller 97
Domains 64, 97, 154, 155

DOT (Department of Transportation) 29, 30, 32, 33, 38, 41, 56, 202, 204

Downlink 24, 35, 86, 102, 103, 115, 140, 204

Download 37, 162, 168, 171, 176, 177

Driver assistance applications 106, 164

Driver assistance systems 3, 18, 157, 165

Driver warning 30, 47, 53

Drivers 2, 4, 5, 6, 18, 73, 80, 156, 157

Drivers and passengers 2, 19, 20, 116, 151, 152, 153, 155, 162

DSPs (digital signal processors) 172, 173, 200

DSRC and LTE C-V2X 83, 200

DSRC Radio Testbed 150

Dynamic data 69, 110, 114, 128

E

EATA (European Automotive and Telecom Alliance) 33, 34, 117, 204

Ecosystem stakeholders 3, 28, 29, 44, 51, 55, 64, 109

ECUs (electronic control units) 12, 73, 96, 97, 99, 100, 106, 160, 196

Edge cloud 72

E/E systems 78

EHorizon 43, 86, 93, 101, 110

Electric power steering (EPS) 12, 96, 205

Electronic horizon 46, 108, 109, 110, 115

Electronic payment 30

Electronic stability control (ESC) 12, 91, 156, 161, 205

Emergency 30, 32, 33, 36, 37, 38, 66, 67, 132

Emergency electronic brake light (EEBL) 30, 32, 33, 36, 37, 38, 52, 66, 67

Emergency vehicle 30, 31, 36, 45, 55, 128

Emergency vehicle notifications 7, 43

Emergency vehicle warning 32, 41, 44, 67
– approaching 30, 35, 38, 52

Enabler, in-vehicle technology 117, 118

ENB 9, 140, 141, 142, 144

Engine 11, 12, 91, 100, 160

Entertainment 28, 151, 152, 158

Entities 168, 170, 171, 174, 177, 178, 189, 193

Environment 4, 5, 38, 39, 41, 86, 89, 94, 110
– surrounding 63, 92

EPS (electric power steering) 12, 96, 205

Equipment 18, 78, 101, 171, 173, 188, 189, 192, 193
– roadside communications 5, 121

Equipment manufacturer 189, 192

Ericsson 35, 36, 37, 44, 87

Error 113, 197, 199, 201

ERTRAC (European Road Transport Research Advisory Council) 25

Essential requirements 170, 192

Ethernet 100, 106, 107, 110, 158, 159, 160, 197, 203

Ethics 17, 19, 21, 195

ETNO European Telecommunications Network Operators' Association ETSI European 205

ETSI 32, 49, 50, 56, 117, 168, 172, 180, 184

ETSI Reconfiguration Ecosystem 174, 175, 177, 179

ETSI security framework for software reconfiguration 188, 189

ETSI software reconfiguration solution 168, 169, 180, 187

ETSI TC 77, 78

European commission 117, 128, 149, 171, 194

European ITS-G5 system 121, 127

European road transport research 3

European Road Transport Research Advisory Council (ERTRAC) 25

E-UTRAN 56, 104, 105

Evolution 23, 24, 25, 48, 49, 54, 55, 83, 96

Exchange 6, 8, 44, 66, 73, 95, 96, 114, 115

Execution 172, 174, 178, 184, 193

Extended sensors 145, 146

External networks 134

F

Failures 48, 54, 102

FCC (Federal Communications Commission) 81, 86, 88, 100

FDM (frequency division multiplexing) 24

Federal communications commission (FCC) 81, 86, 88, 100

FHWA (Federal Highway Administration) 29, 56, 86, 100, 122, 124, 205

Field programmable gate arrays (FPGAs) 106, 200

Firmware 48, 71, 107, 109, 114, 115

Fleet 3, 64, 65, 77

FlexRay 47, 97, 99, 106, 107, 159, 160, 197
Flow controller 174, 177, 178, 205
Footprint analysis report 124, 125
Forward collision warning (FCW) 7, 11, 33, 38, 41, 43, 49, 52, 66
Forward collision warning, cooperative 31, 32, 36, 52
FOTA 86, 90, 99, 114, 115
FPGAs (field programmable gate arrays) 106, 200
Framework, radio control 174, 175, 177, 178, 179
Frequency bands 23, 80, 119, 126, 131, 186
Front-end 175, 176, 182
Functional blocks 177
Functional safety 15, 78, 79, 80, 86, 91, 111, 116, 196
Functionalities 112, 114, 127, 155, 162, 163, 178, 196
Functions 15, 73, 80, 86, 97, 163, 164, 178, 196
– generic 128

G
Generation 12, 14, 40, 41, 128, 132, 193, 203, 209
Germans 4, 18, 204
Germany's drivers 19
GHz 74, 81, 82, 84, 100, 125, 126, 147, 148
GHz band 49, 81, 82, 84, 85, 100, 101, 124, 201
Gigabytes 60, 65, 86, 97, 130
Global system 132, 205
Global system for mobile (GSM) 132, 205
GLOSA (green light optimized speed advice) 42, 101
GNSS 8, 10, 89, 110, 141, 142, 147, 160
GPS (global positioning systems) 123, 151, 152, 155, 157
GPUs (graphic processing units) 48, 98, 106, 205
Graphic processing units (GPUs) 48, 98, 106, 205
Green light optimized speed advice (GLOSA) 42, 101
GSM (Global System for Mobile) 132, 205

H
Hardware 15, 16, 89, 90, 106, 181, 191, 192, 193
Hardware and software 170, 192, 195, 196
Hardware components 96, 192, 193
Hardware manufacturer 170, 171, 191, 193
Hardware platforms 90, 180
HAV (highly automated vehicles) 50, 51, 205
HD maps 22, 90, 95
– dynamic 12, 69, 70, 108
Head-up displays (HUD) 42, 44, 156, 157
HEV (hybrid electric vehicle) 96, 205
High definition (HD) 10, 12, 37, 44, 50, 69, 108, 205
Highly automated vehicles (HAV) 50, 51, 205
Highways 5, 6, 15, 22, 45, 55, 61, 101, 114
HMI systems 151, 162, 163
HMIs (human-machine-interface) 12, 46, 86, 151, 152, 154, 155, 156, 157
Https 25, 26, 57, 87, 88
Huawei 35, 36, 37, 41, 42, 56, 87
Human driver 17, 22, 62, 73, 113
HW Radio platform 175, 176
Hybrid communication 33, 34
Hybrid electric vehicle (HEV) 96, 205

I
ICA (intersection collision avoidance) 66, 101
IEEE (Institute of Electrical and Electronic Engineering) 14, 74, 75, 121, 125, 149, 198, 200, 201
IEEE 802.11p 74, 75, 76, 77, 81, 82, 83, 88, 133
IEEE 802.11p DSRC 83, 87
IEEE Intelligent Vehicles Symposium 116, 118
IEEE International Conference 149, 201
IEEE standard 87, 88
IEEE standard for wireless access in vehicular environments 87, 88
IEEE Veh 150
IEEE Vehicular Networking Conference 116, 117
IETF RFC 75, 88
Implementation 109, 111, 112, 113, 115, 173, 174, 186, 187
IMT 82, 129

Information 2, 6, 8, 13, 39, 56, 145, 149, 170
Information technology 2, 87, 88, 194
Infotainment 37, 38, 86, 90, 91, 152, 158, 162, 164
Infotainment systems 16, 24, 154, 155, 156, 162, 163, 164
– current 24, 156
Infrastructure 8, 9, 20, 21, 34, 43, 64, 78, 102
– cellular 10, 77, 131, 136
– roadside 9, 47, 49, 87, 100
– surrounding 197
Initialization 182, 183
Input 39, 44, 108, 110, 152, 154, 156, 163, 182
Installation 167, 169, 173, 184, 185, 189, 193
Instrument cluster 42, 157
Insurance 21, 24, 28, 37, 38, 47, 50, 52, 54
– based vehicle 47
Integration 17, 37, 42, 54, 66, 152, 161, 162, 164
Intelligent road transport systems 56, 202
Intelligent transport systems 25, 56, 64, 66, 87, 118, 125
– cooperative 38, 150
Intelligent transportation systems (ITSC) 6, 47, 54, 55, 88, 149, 201, 204, 206
Intel's new autonomous vehicle center 118
Interact 1, 19, 151, 155, 174
Internet 20, 22, 27, 28, 156, 158, 160, 161, 162
Internet access 62, 68, 151, 155, 158, 160
Internet Engineering Task Force (IETF) 74
Intersection collision warning (ICW) 32, 36, 43, 52, 54, 55, 66, 205
Intersection movement assist (IMA) 33, 37, 38, 44, 46, 52, 54, 206
Intersections 37, 46, 47, 49, 50, 53, 72, 108
In-vehicle 59, 72, 86, 152, 161, 196
In-vehicle ADAS/AD systems 44
In-vehicle commerce 38
In-vehicle connectivity 46, 161
In-vehicle infotainment (IVI) 16, 50, 51, 71, 101, 152, 153, 154, 155, 160, 161, 162, 163, 164
In-vehicle infotainment systems 153, 154, 155, 161, 162, 164

In-vehicle networking and connectivity next-generation technologies 100
In-vehicle networks 12, 73, 91, 100, 158, 197
In-vehicle sensors 11, 43, 47, 54, 86, 91
In-vehicle signage 32, 35
IP (Internet Protocol) 75, 88, 187, 206

ISMVL (International Symposium on Multiple-Valued Logic) 116, 118
ISO/TR 38, 56
ITS-G5C 126
IVI systems 153, 154, 155, 163, 164

J
JTNC (Joint Tactical Networking Center) 187, 194

L
Lane 43, 44, 93, 94, 95, 108, 109, 110, 111
Lane assist 43, 66, 91, 110, 111
Lane change assist 44, 47, 52
Lane change warning (LCW) 31, 33, 36, 38, 46, 54
Lane departure warning (LDW) 8, 11, 43, 44, 52, 93, 95, 157, 206
Lane keep assist (LKA) 52, 98, 109
Latency 35, 36, 63, 66, 67, 95, 98, 135, 158
– low 13, 45, 49, 50, 59, 85, 130, 161
Latency in milliseconds 36, 66, 67
Latency reduction 144, 146
Laws 80, 112, 164, 194
Layers 47, 69, 70, 75, 99, 108, 126, 206, 207
LDM (Local Dynamic Map) 47, 127, 128
Left turn assist (LTA) 36, 37, 38, 45, 46, 52, 206
Levels 4, 8, 10, 17, 54, 55, 73, 74, 80
LIDAR 10, 11, 12, 89, 90, 94, 95, 96, 113
LIDAR sensors 49, 60, 92, 94, 98, 109, 158
LIN (local interconnect network) 47, 99, 100, 106, 159, 160, 197, 206
Link 6, 38, 47, 80, 100, 111, 112, 135, 159
LKA (lane keep assist) 52, 98, 109
LLC (logical link control) 75, 88
Local area 30, 31, 32
Localization 48, 69, 90, 106, 108
Location 10, 43, 48, 50, 63, 65, 107, 108, 112
Logical link control (LLC) 75, 88

Low-level parameters 190, 191
Low-voltage differential signalling (LVDS)
    159
LTE 14, 15, 49, 50, 68, 69, 103, 131, 135
LTE C-V2X 74, 75, 82, 83, 167, 179, 198, 199,
    200
LTE Networks 9, 102
LTE V2X 76, 121, 135, 200
LTE Vehicular Services 131, 133, 135, 137,
    139, 141

M
MAC (medium access control) 14, 75, 77, 87
Maintenance 21, 38, 50, 86, 133, 134, 160,
    178, 207
Management 9, 39, 75, 77, 100, 124, 125,
    127, 178
Manufacturers 71, 75, 169, 170, 171, 173,
    184, 189, 193
− incumbent vehicle 28
− mobile device 155, 177
− module 186
− premium vehicle 155, 195
Map updates 108, 195
Maps 14, 43, 44, 69, 70, 108, 109, 110, 115
Market 75, 83, 85, 171, 192, 197, 198, 199,
    200
Market penetration 105, 198, 199, 200
Martin 118, 165, 194
MCU (micro controller units) 96, 107, 152,
    206
MDRC (mobile device reconfiguration clas-
    ses) 172, 173, 178, 206
Media 99, 159, 160, 161
Memory, program 183
Message sublayer 77, 179
Messaging 20, 66
Meters 35, 36, 50, 66, 69, 92, 93, 94, 110
MHz 43, 74, 81, 82, 148
Middleware 107, 169, 180, 181, 187, 193
Mobile broadband 38, 40, 46, 47
Mobile devices 9, 63, 154, 155, 156, 185,
    187, 194, 206
− reconfigurable 174, 177, 185
Mobile network operators (MNOs) 39, 41,
    82, 85, 129
Mobile smartphone 161
Mobility 8, 9, 25, 27, 28, 64, 66, 100, 195
Mobility services 24, 27, 28

Modes 55, 61, 62, 110, 140, 144, 157
− device-to-device communication 131, 132
Monitor 4, 5, 28, 73, 74, 91, 177, 178
Motorways 4, 5
MPM (mobility policy manager) 177, 178,
    185
MRC (Multiradio Controller) 174, 178
MURI (Multi Radio Interface) 174, 175, 176,
    177, 178, 184, 185, 186, 194

N
Navigation 31, 32, 46, 47, 90, 91, 107, 108,
    156
− augmented 38, 46, 47
Navigation and telematics 12, 89, 110, 112,
    113, 115
Navigation control unit (NCU) 152
Navigation systems 112, 152, 154, 157, 161
NCU (navigation control unit) 152
Network assistance 49
Network function virtualization (NFV) 50
Network layer 75, 77
Network operators 82, 185, 189
Network slicing 34, 40
Network topologies 100, 107
Networked autonomous vehicles 20, 21, 22
Networked vehicle system 15, 24
Networking 13, 55, 63, 64, 66, 81, 82, 114,
    196
− cellular 16, 17, 23, 201
− wireline 158, 159
Networking and connectivity technology 55,
    63, 71, 89, 161
Networking stack 177, 178, 185
Networking technologies 3, 99
Networks 6, 8, 9, 14, 15, 77, 78, 100, 101
− cellular 49, 131, 161, 165
− controller area 99, 159, 204
− local interconnect 99, 159
− metropolitan area 87, 88
− social 2, 28
− vehicular 114, 117, 150, 165
Neural networks 106, 116, 118
− recurrent 99, 106
New SW component 192
NFC (near field communication) 59, 86, 100,
    107, 112, 153, 206
NFV (network function virtualization) 50
NGMN 37, 56

NHTSA (National Highway Traffic Safety Administration) 14, 22, 80, 86, 124, 125, 202
NJ 116, 117, 118, 149, 201
NR (New Radio) 68, 69, 73, 87, 105, 131, 146, 197, 199
NS-2 Simulation Platform 150

O
On-Road Motor Vehicle Automated Driving Systems 26, 88
Open system interconnect (OSI) 75
Operating systems and communications drivers and applications 113
Operation 104, 105, 126, 169, 171, 174, 181, 182, 183
Operators 21, 41, 80, 181, 182, 183, 192
Option 64, 81, 82, 111, 114, 151, 152, 159, 161
OSI (open system interconnect) 75
Outlook 150, 195, 196, 198, 200, 202

P
Park assist 11, 47, 110, 111
Passengers 2, 20, 27, 28, 72, 116, 151, 152, 153
Path planning 48, 108, 109
PCW (pre-crash-warning) 43, 66
PD (pedestrian detection) 8, 93, 207
PDR (packet delivery ratio) 63
Pedestrian detection (PD) 8, 93, 207
Periodic broadcast 67, 68, 115
Periodic broadcast vehicle-mode 67
Phase 16, 34, 130, 132, 143, 144, 145, 147, 149
Piscataway 116, 149, 201
PKI (public key infrastructure) 13, 78, 207
Platform 48, 167, 169, 172, 173, 179, 181, 188, 189
Platoon 10, 32, 37, 39, 49, 145
Platooning 9, 15, 37, 44, 50, 74, 77, 145, 146
Politics 17, 19, 21, 195
Positioning 43, 47, 50, 71, 90, 109, 110, 146, 147
Positioning accuracy 35, 36, 50, 146
Positioning systems, global 151, 152, 157
Pre-crash-warning (PCW) 43, 66
Pressure 11, 83, 91

Priority 20, 21, 142, 145, 201
Products 18, 27, 75, 83, 85, 193
Proof 15, 186, 189
Protocols 16, 107, 115, 121, 134, 139, 160, 194
– transmission control 75, 88
PSCCH 140, 148
PSSCH 140, 148
Public key infrastructure (PKI) 13, 78, 207

Q
QoS (quality of service) 101, 121, 158, 167
Qualcomm 23, 35, 36, 37, 41, 57, 87, 88

R
RA code 180
Radar 10, 11, 43, 89, 90, 93, 95, 98, 113
Radar sensors 60, 92, 93, 94
Radio 20, 173, 174, 175, 176, 181, 182, 187, 189
Radio access network working group 134
Radio application (RA) 173, 174, 177, 178, 185, 188, 189, 190, 207
Radio application developer 189, 190, 191
Radio application package 177, 182, 188
Radio applications 174, 177, 180, 181, 188, 189, 190
– unified 174, 175, 177, 185, 186, 190
Radio Apps 175, 176
Radio behavior 169, 171, 190
Radio characteristics 170, 171, 174, 187
Radio communication 186
Radio computer 174, 175, 176, 178, 185
Radio Connection Manager (RCM) 174, 177, 178, 207
Radio data system (RDS) 112, 151
Radio equipment 170, 171, 188, 189, 190, 194
Radio equipment directive 169, 170, 171, 190, 192
Radio interfaces 136, 146
Radio Layer 134
Radio library 175, 176, 181, 182, 183
Radio parameters 170, 178, 191
Radio platform 174, 175, 176, 177, 189, 191
Radio platform driver 174
Radio Programming Interface (RPI) 175, 176, 177, 194
Radio resources 24, 103, 178

RadioApp 168, 174, 177, 188, 192
Rain 46, 92, 93
Range 11, 13, 35, 38, 43, 46, 92, 94, 154
– functional 156, 163
Ranging 93, 94, 96, 114, 163, 206
RAP 183
RAs 178
RATs (radio access technologies) 39, 153,
    169, 172, 173, 177, 178, 185, 193
RCF (radio control framework) 174, 175, 177,
    178, 179, 185, 207
RCTA (rear cross traffic alert) 95, 207
RDS (radio data system) 112, 151
Rear 43, 47, 50, 93, 95, 153
Rear cameras 111, 158
Rear cross traffic alert (RCTA) 95, 207
Received signal strength indication (RSSI)
    63, 208
Recognition, traffic sign 11, 42, 43, 53, 95,
    118
Reconfigurability 169, 172, 190, 191
Reconfigurable radio frequency interface.
    See RRFI
Reconfigurable Radio Systems 167, 194
Reconfiguration 15, 171, 172, 185, 188, 190,
    191
Reconfiguration policy 188, 189
Recurrent neural networks (RNN) 99, 106,
    207
Red light violation warning (RLVW) 33, 35,
    53, 207
Regulation administrations 82, 84, 85
Regulation Framework 169, 171
Regulations 21, 22, 23, 24, 25, 59, 60, 80,
    86
Remote attestation 188, 189
Replacement 84, 169, 184, 193
Request 2, 3, 5, 24, 99, 100, 176, 177, 191
Resolution 96, 109, 111, 113, 114
Resource manager 173, 174, 177, 178, 207
Responsibility 20, 21, 133, 170, 191, 192,
    193, 195
RF Transceiver 174, 177
RLVW (red light violation warning) 33, 35,
    53, 207
RNN (recurrent neural networks) 99, 106,
    207
Road 1, 2, 6, 27, 61, 72, 94, 98, 111
– public 22, 64, 65, 79

Road hazards 37, 46, 49, 69, 70, 72
Road network 43, 78
Road safety 34, 36, 39, 51, 52, 53, 62, 77, 81
Road safety services 39, 41
Road side unit 137, 138, 147, 149
Road traffic 18, 20, 22, 43
Road traffic acts 30, 51, 52, 53, 54, 80
Road users, vulnerable 34, 36, 37, 46, 77
Roadmap 22, 25, 26, 27
RPI (Radio Programming Interface) 175, 176,
    177, 194
RRFI (reconfigurable radio frequency inter-
    face) 177, 186, 194, 208
RRS (reconfigurable radio systems) 167,
    189, 192, 194, 204
RSSI (received signal strength indication)
    63, 208
RSUs (road side units) 9, 104, 105, 107, 137,
    138, 145, 147, 149
RVM (Radio Virtual Machine) 175, 180, 181,
    182, 183, 187, 190, 191, 194
RVM classes 190
– highest-level 190, 191
– intermediate-level 191
– lowest-level 190, 191
RVM software component 191

S
SA1, 133, 134, 135, 145
SAE (Society of Automotive Engineers) 3, 4,
    14, 22, 34, 73, 74, 87, 101
SAE level 62, 90, 92, 96, 98, 106, 109, 110,
    195
Safety 13, 14, 20, 21, 42, 64, 82, 125, 158
Safety applications 7, 14, 32, 122
Safety data 44, 158
Safety requirements 12, 49, 91
– functional 86, 92
Safety-critical functions 73, 74, 80
SCA (Software Communications Architec-
    ture) 169, 187, 194
Scenarios and use cases 42, 46, 48
Scenes 61, 72
SDOs (standards developing organizations)
    39, 59
SDRs (software-defined radio) 187, 194
Security 34, 35, 77, 86, 112, 113, 115, 116,
    194
Security requirements 35, 54, 194

Selected MAC control services 126

Selected MAC data services 126

Self-driving vehicles 3, 7, 8, 15, 17, 25, 27, 64, 65

– fleet of 61, 65

Sensor data 10, 12, 47, 48, 49, 94, 96, 98, 99

Sensor data fusion 49, 94, 98, 195

Sensor extensions 62, 70, 86, 114

Sensors 10, 11, 17, 24, 59, 65, 89, 91, 114

– local 145

Sensors and actors 10, 23, 91, 106

Servers 8, 12, 105, 106, 109, 138, 139, 145, 195

Service providers 27, 34, 157, 192

Services 9, 16, 28, 52, 55, 56, 85, 178, 185

– based 50, 82, 107

SFBs (standard functional blocks) 182, 183

Sharing 10, 37, 38, 39, 45, 49, 114, 124, 144

Signals 7, 42, 70, 97, 124, 125, 141

Situational awareness 24, 37, 46, 114

Situations 2, 4, 5, 21, 22, 83, 84, 190, 191

Smartphone apps 162, 163

Smartphone integration 59, 152, 153, 161

Smartphones 112, 114, 151, 152, 153, 154, 161, 162, 165

Snow 46, 92, 93

Society 3, 17, 19, 20, 21, 33, 34, 36, 74

Software 16, 90, 105, 113, 169, 170, 171, 192, 196

Software applications 115, 196

Software communications architecture specification 187, 194

Software components 73, 75, 167, 168, 169, 171, 172, 173, 174

– reconfigured 190

Software developers, third-party 184

Software development 106, 196

Software features 15

Software management, based vehicle 90

Software manufacturer 191, 192, 193

Software reconfigurability 167, 169, 170

Software reconfiguration 167, 168, 172, 173, 174, 176, 179, 180, 188

Software updates 37, 39, 71, 107, 109, 157, 173

Software-defined radio (SDRs) 187, 194

Solutions 3, 16, 24, 44, 48, 80, 132, 160, 187

SOS services 30, 32

SOTA 86, 90, 107, 109, 114, 115

SOTA updates 86, 109

SPaT (signal phase and timing) 14, 16, 101, 121

Specifications 77, 78, 101, 133, 134

Spectrum 29, 82, 85, 87, 100, 103, 124, 126

Spectrum band 81, 82

Spectrum efficiency 129, 131

Speech dialog system 154

Speeds 14, 15, 131, 195

Springer 116, 117, 165

Stage, early 184, 186, 195, 198, 200

Standardization 22, 36, 82, 86, 87, 88, 106, 124, 206

Standards 59, 60, 74, 75, 77, 78, 82, 87, 111

Standards developing organizations (SDOs) 39, 59

State 5, 20, 25, 26, 27, 37, 118, 124, 182

Stations 56, 126, 128

Storage 54, 55, 72, 73, 89, 90, 96, 175, 176

Strategy analytics 118, 165

Suppliers 25, 27, 41, 44, 45, 47, 112, 117, 164

– wireless infrastructure 49, 50

System architecture for radio computer 175, 176

System architecture working group 133

System components 6, 8, 15

System malfunctions 82, 111, 196

System requirements 82, 127

System-on-chips (SoC) 187

Systems 4, 5, 20, 22, 73, 74, 84, 86, 111

T

Tablets 151, 152, 153, 154, 161, 165

Target platform 168, 169, 172, 180, 183, 184

Task 4, 21, 55, 62, 64, 155

TB 3, 208

TC (Technical Committee) 77, 192, 208

TCP (transmission control protocol) 75, 88, 179, 208

TCU (transmission control units) 13, 44, 47, 152

Technical committee (TC) 77, 192, 208

Technical specifications. *See* TS

Technologies 1, 6, 55, 88, 89, 96, 116, 124, 200

– sensor 90, 92, 95, 195

TEE (trusted execution environments) 115
Telematics 13, 90, 91, 112, 151, 152, 162, 163, 164
Telematics and control 152, 156, 157, 163, 164
Telematics Industry Application Alliance (TIAA) 38
Telematics services 50, 156, 157
Terminals working group 133, 134
TIAA (Telematics Industry Application Alliance) 38
Time of flight (ToF) 94, 149, 208
Time sensitive networking (TSN) 159, 209
Timeline 130, 131, 132
Tire pressure monitoring system (TPMS) 96, 112
TJW (traffic jam warning) 43, 66, 208
TLI (traffic light information) 42, 208
TMC (traffic message channel) 91, 151, 156
Toll collection, electronic 32, 52, 66, 68, 91, 205
Torque 12, 91, 110
TPMS (tire pressure monitoring system) 96, 112
TR (Technical Reports) 39, 40, 41, 56, 134, 135, 192, 194, 208
Traffic 2, 21, 22, 24, 27, 28, 37, 42, 43
Traffic congestion 13, 69, 70, 197
Traffic data 30, 108, 110, 156
Traffic jam assist 10, 42, 44, 53, 109
Traffic jam warning (TJW) 43, 66, 208
Traffic light information (TLI) 42, 208
Traffic management 39, 47, 50, 68
Traffic message channel (TMC) 91, 151, 156
Traffic participants 44, 47, 80, 94, 95
Traffic sign recognition (TSR) 11, 42, 43, 53, 93, 95, 118, 209
Traffic signals 30, 108
Transit vehicle data transfer 31
Transmission 9, 16, 100, 102, 140, 142, 158
Transmission control units (TCU) 13, 44, 47, 152
Transport 3, 9, 27, 33, 34, 38, 46, 78, 86
Transverse 4, 5
Trials 15, 37, 41, 44, 45, 46, 76
Trusted execution environments (TEE) 115
TS (Technical Specifications) 39, 56, 134, 135, 168, 188, 194, 208
TSN (time sensitive networking) 159, 209

TTS (Traffic Technology Services) 42

U
UDFB Set 183
UDFBs (user-defined functional blocks) 182, 183
UDP (user datagram protocol) 75, 88, 179, 209
UE (user equipment) 9, 44, 134, 136, 140, 141, 142, 147, 209
Ultrasonic sound 90, 95, 109, 113, 195
Ultrasonic sound sensors 90, 92, 93, 98
United States 22, 25, 38, 41, 43, 79, 81, 86, 195
Units 9, 90, 112, 153
– vehicle head 107, 153, 160
Uplink 24, 35, 86, 102, 103, 115, 140, 209
Upload 89, 109, 152, 175, 176, 193
URA (unified radio applications) 174, 177, 178, 185, 186
URAI (Unified Radio Application Interface) 174, 175, 176, 177, 178
URLLC sidelink 146
US 81, 84, 150
USB 107, 152, 154
User interface, graphical 153, 154, 155
User interfaces (UI) 20, 56, 118, 153, 154, 155, 157, 165, 209
Users 75, 78, 121, 123, 155, 168, 171, 188, 189
Uu interface 104, 105

V
V2X 59, 60, 62, 64, 66, 76, 77, 78, 146
VCU (vehicle control unit) 152
Vehicle acts 10, 109
Vehicle communications 77, 167, 198
Vehicle components 100, 163
Vehicle computing platforms 13, 106, 196
Vehicle connectivity 44, 45
Vehicle control 21, 22, 86, 90, 91, 106, 107, 112, 113
Vehicle control system 22, 98
Vehicle control unit (VCU) 152
Vehicle diagnostics 24, 152, 157
Vehicle drivers 3, 21, 22, 27, 152, 161, 164
Vehicle drivers and passengers 3, 27, 28, 152, 153, 161, 164
Vehicle dynamics 24, 56, 165

Vehicle ecosystem 27, 28, 90, 107, 115, 116, 195, 196
Vehicle ecosystem infrastructure 112, 196
Vehicle fleets 163, 195
Vehicle functions 15, 91, 96, 153
Vehicle HMIs 151, 153, 162
Vehicle industry stakeholders 27, 63
Vehicle infrastructure 55, 65, 71, 72, 100, 195, 196
Vehicle infrastructure integration 33, 56
Vehicle maintenance 37, 50, 112
Vehicle manufacturers 3, 41, 42, 53, 106, 107, 153, 157, 164
Vehicle models 156, 164
Vehicle navigation 90, 108
Vehicle navigation systems 12, 108, 115
Vehicle networking 2, 3, 7, 8, 9, 15, 16, 22, 23
Vehicle operation 15, 16, 54
Vehicle positioning 146
Vehicle radio 151
Vehicle regulations 21, 22
Vehicle safety communications 56, 118
Vehicle sensors 36, 39, 65, 69, 91, 92, 93, 96, 152
Vehicle surroundings 59, 72, 108
Vehicle system components 71, 152
Vehicle systems, autonomous 3, 105
Vehicle tests 12
Vehicles 1, 17, 43, 70, 72, 89, 94, 101, 112
– connecting 9, 10, 39
– driverless 18
– host 65, 66, 75
– remote 66, 75, 145
– software in 90, 105
– surrounding 69, 70, 93, 94, 108, 132
Vehicle-to-cloud 197, 201
Vehicle-to-device 66
Vehicle-to-everything 6, 9, 10, 25, 129, 135, 209
Vehicle-to-home 29, 54, 66, 209
Vehicle-to-infrastructure 6, 8, 9, 10, 38, 39, 136, 138, 139
Vehicle-to-Infrastructure Communications 122, 137
Vehicle-to-network 9, 10, 29, 39, 85, 136, 139, 209
Vehicle-to-network communication 138, 139
Vehicle-to-pedestrian 9, 10, 39, 49, 66, 138

Vehicle-to-person 85, 136, 137, 209
Vehicle-to-vehicle 8, 9, 10, 14, 38, 39, 77, 136, 137
Vehicle-to-vehicle communication 22, 131, 138
Vehicular communications 25, 56, 59, 77, 82, 83, 87, 101, 117
Vehicular environments, wireless access in 74, 75, 87, 88, 125
Vehicular safety communications 150
Vehicular services 85, 132, 133, 137, 198
– works on LTE 131, 133, 135, 137, 139, 141
Velocity, relative inter-vehicle 114
Video 28, 32, 37, 46, 49, 91, 95, 153, 162
Video camera sensors 93
Video camera systems 10, 90
Video cameras 11, 89, 90, 92, 94, 96, 98, 113, 115
Video data 12, 39, 72, 73
Virtual machine 181, 187, 190, 194
Virtual reality (VR) 37, 47, 72, 209
Visibility enhancer 30
VNC (Vehicular Networking Conference) 116, 117
Vodafone 35, 37, 87, 157
Voice control 152, 156
VR (virtual reality) 37, 47, 72, 209
Vulnerabilities 112, 115, 167, 168
Vulnerable road user (VRU) 32, 34, 35, 36, 37, 46, 48, 50, 67

W, X, Y, Z
Warning 9, 10, 38, 50, 51, 53, 66, 67, 68
– blind spot 31, 33, 38, 52
– intersection collision 32, 36, 43, 54, 55, 66
– overtaking vehicle 32, 67
– queue 7, 37, 41, 44, 46, 53, 68
– road condition 31, 38
– speed 7, 37, 38, 41, 44, 109
– stationary vehicle 32, 43, 66, 67
WAVE (Wireless Access in Vehicular Environments) 13, 14, 74, 75, 87, 88, 94, 100, 101
Weather 16, 28, 43, 47, 50, 54, 55, 110
Weather conditions 29, 65, 93, 98, 113
Web services 156
Wheels, steering 1, 2, 28, 42, 46, 73, 74, 156, 157

WI 77, 133, 135, 144, 209
Wi-Fi 59, 152, 153, 157, 158, 162, 165, 190,
     191
WiGig channels 84, 85
Wireless networking 17, 112, 115, 116, 153,
     160, 161, 201
Wireless networking and connectivity 151,
     161, 162, 163, 201

Wireless technologies 13, 17, 78, 101, 116
Wireless V2X networking 100, 114
Wireless V2X networking and connectivity
     technologies 100
Working Items (WIs) 77, 133
WSMP (WAVE short message protocol) 14,
     75